Elementary Probability for Application:

This book is an introduction to pro
most useful for applications. Its philo
in action and so there are more than 3
several classics such as the birthday prc
of applications not found in other bool

Rick Durrett received his Ph.D. in ope
After 9 years at UCLA, he came to Cori
mathematics. He is the author of 8 books a
of topics, and he has supervised more than
Academy of Science and the American Ac
Institute of Mathematical Statistics.

Elementary Probability for Applications

Rick Durrett

Department of Mathematics, Cornell University

CAMBRIDGE UNIVERSITY PRESS
Cambridge, New York, Melbourne, Madrid, Cape Town, Singapore,
São Paulo, Delhi, Dubai, Tokyo

Cambridge University Press
32 Avenue of the Americas, New York, NY 10013-2473, USA

www.cambridge.org
Information on this title: www.cambridge.org/9780521867566

First published 2009

Printed in the United States of America

A catalog record for this publication is available from the British Library.

Library of Congress Cataloging in Publication data

Durrett, Richard, 1951–
Elementary probability for applications / Rick Durrett.
p. cm.
Includes bibliographical references and index.
ISBN 978-0-521-86756-6 (hardback)
1. Probabilities. I. Title.
QA273.D8638 2009
519.2–dc22 2009020688

ISBN 978-0-521-86756-6 Hardback

Contents

Preface

Probability is the most important concept in modern science especially as nobody has the slightest notion what it means.

BERTRAND RUSSELL

About 15 years ago, I wrote the book *Essentials of Probability*, which was designed for a one-semester course in probability that was taken by math majors and students from other departments. This book is, in some sense, the second edition of that book, but there are several important changes:

- Chapter 1 quickly introduces the notions of independence, distribution, and expected value, which previously made their entrance in Chapters 2, 3, and 4. This makes it easier to discuss examples; for example, we can now talk about the expected value of bets.
- For 5 years these notes were used in a course for students who knew only a little calculus and were looking to satisfy their distribution requirement in mathematics, so it is aimed at a wider audience.
- Markov chains are covered, and thanks to a suggestion of Lea Popovic, this topic appears right after the notion of conditional probability is discussed. This material is usually covered in an undergraduate stochastic processes course, if you are fortunate enough to offer one in your department, but in our experience this material is popular with students.
- Continuous distributions are presented as an optional topic. This decision originated to minimize the reliance on calculus, but in time I have grown to enjoy abandoning the boring mechanics of marginal and conditional distributions to spend more time talking about probability.

This book, like its predecessor, takes the philosophy that the best way to learn probability is to see it in action. There are more than 350 problems and 200 examples. These contain all the old standards: the World Series, dice and card games, birthday problem, Monty Hall, medical testing posterior probabilities, and various applications of the central limit theorem. However, it also contains a number of topics that are rarely covered: Benford's law, the TV show

Deal or No Deal, alliteration in Shakespeare, Wayne Gretzky's scoring record, lottery double winners, the trials of O. J. Simpson and Sally Clark, how to play blackjack, cognitive dissonance in monkeys, the hot hand in basketball, and option pricing.

The discussion in the text follows the philosophy of *Dragnet's* Joe Friday: "Just the facts, ma'am." However, there are several dozen figures that show the shapes of distributions and help to explain the arguments. Copyright laws prevent me from calling this book "Probability for Dummies," but that would be a misnomer. The first two words of the title are "Elementary Probability," which means that there are few formal prerequisites. I have done my best to explain things clearly, but the reader (and the instructor) should be warned that thinking is required.

How should you teach from this book?

The easiest way to explain this is with a flowchart of the chapters:

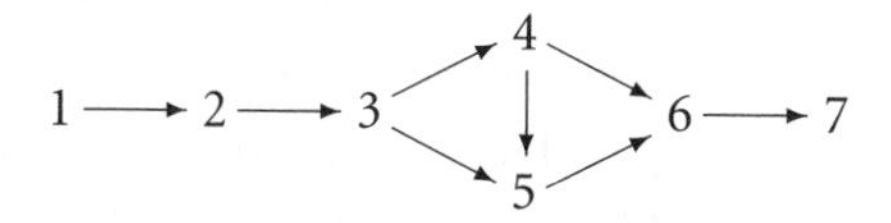

Chapter 1 introduces the language we need to talk about examples. Chapter 2 covers combinatorial probability and is followed by Chapter 3's treatment of conditional probability. At this point, one can go on to Markov chains in Chapter 4 (which I prefer) or to continuous distributions in Chapter 5. From Chapter 4 you can go to Chapter 5 or leave this boring topic to the instructor of the statistics course that follows yours and go on to the law of large numbers and central limit theorem in Chapter 6. The all-important normal is a continuous distribution, of course, but all computations for it are done with tables, so the only concept one needs is the distribution function. Finally, Chapter 7 is a brief introduction to option pricing. I find this makes a nice final lecture before one turns to the business of reviewing material in preparation for the final exam.

Supporting cast

The writing of this book benefited from the comments of several people paid by Cambridge University Press to read various chapters, and in particular by the efforts of one reader who made hundreds of comments on the writing style. In the spring quarter of 2008, Ed Waymire used the book at Oregon State, and in the fall quarter of 2008, Michael Phelan used the book at U. C. Irvine. I am grateful to Michael for his many comments and his enthusiasm for the book.

When *Essentials of Probability* was written, my sons David and Greg were 6 and 4, and Tipper Gore was complaining about the lack of values in the music I listened to. Now David is a senior at Ithaca College, one semester away from graduating with a major in journalism, and wondering if the economic collapse brought on by 8 years of the Bush administration will keep him from getting a job. Greg, who is a junior at MIT double majoring in computer science and math, has better long-term job prospects, but he will probably go to graduate school before deciding how close he wants to be to the real world.

This brings me to the two women who are the most important for this book. The first is my wife, Susan. After 28 years of marriage and almost a dozen prefaces, I have run out of clever things to say. When the kids are home from college, as they are now during winter break, she is a flurry of activity. In between, she fills her empty nest with the *New York Times*, its crossword puzzles (an addiction I share), and tending to her parents who moved to Ithaca from the Sacramento area about 5 years ago. In December, we had our first vacation away together since David was born in 1988. Before you say "how romantic," I should admit that the trip consisted of 2 days on the beach at Half Moon Bay and 3 days of work in Berkeley.

The other important woman is my editor Lauren Cowles. After seeing *Essentials of Probability* moved around from Wadsworth to Duxbury Press and on to International Thompson Publishing and then go out of print without my being told, it is nice to be in the hands of someone who cares about my books. Even though I (and others) have spent a lot of effort debugging the book, it is inevitable that there will be typos. Email them to rtd1cornell.edu and look for lists on `www.math.cornell.edu/~durrett`, where you can find information about my research and other books.

Rick Durrett

1

Basic Concepts

In this chapter we introduce the basic terminology of probability theory. The notions of independence, distribution, and expected value are studied in more detail later, but it is hard to discuss examples without them, so we introduce them quickly here.

1.1 Outcomes, events, and probability

The subject of probability can be traced back to the 17th century when it arose out of the study of gambling games. As we see, the range of applications extends beyond games into business decisions, insurance, law, medical tests, and the social sciences. The stock market, "the largest casino in the world," cannot do without it. The telephone network, call centers, and airline companies with their randomly fluctuating loads could not have been economically designed without probability theory. To quote Pierre-Simon, marquis de Laplace from several hundred years ago:

> It is remarkable that this science, which originated in the consideration of games of chance, should become the most important object of human knowledge . . . The most important questions of life are, for the most part, really only problems of probability.

In order to address these applications, we need to develop a language for discussing them. Euclidean geometry begins with the notions of point and line. The corresponding basic object of probability is an **experiment**: an activity or procedure that produces distinct, well-defined possibilities called **outcomes**. (Here and throughout the book **boldface type** indicates a term that is being defined.)

Example 1.1 If our experiment is to roll one die then there are 6 outcomes corresponding to the number that shows on the top. The set of all outcomes in this case is $\{1, 2, 3, 4, 5, 6\}$. It is called the **sample space** and is usually denoted by Ω.

Symmetry dictates that all outcomes are equally likely, so each has probability 1/6.

Example 1.2

Things get a little more interesting when we roll two dice. If we suppose, for convenience, that they are red and green then we can write the outcomes of this experiment as (m, n), where m is the number on the red die and n is the number on the green die. To visualize the set of outcomes it is useful to make a small table:

$$\begin{array}{cccccc} (1,1) & (2,1) & (3,1) & (4,1) & (5,1) & (6,1) \\ (1,2) & (2,2) & (3,2) & (4,2) & (5,2) & (6,2) \\ (1,3) & (2,3) & (3,3) & (4,3) & (5,3) & (6,3) \\ (1,4) & (2,4) & (3,4) & (4,4) & (5,4) & (6,4) \\ (1,5) & (2,5) & (3,5) & (4,5) & (5,5) & (6,5) \\ (1,6) & (2,6) & (3,6) & (4,6) & (5,6) & (6,6) \end{array}$$

There are $36 = 6 \cdot 6$ outcomes since there are 6 possible numbers to write in the first slot and for each number written in the first slot there are 6 possibilities for the second.

The goal of probability theory is to compute the probability of various events of interest. Intuitively, an event is a statement about the outcome of an experiment. The formal definition is: An **event** is a subset of the sample space. For example, "the sum of the two dice is 8" translates into the set $A = \{(2, 6), (3, 5), (4, 4), (5, 3), (6, 2)\}$. Since this event contains 5 of the 36 possible outcomes its probability $P(A) = 5/36$.

For a second example, consider B = "there is at least one 6." B consists of the last row and last column of the table, so it contains 11 outcomes and hence has probability $P(B) = 11/36$. In general, the probability of an event C concerning the roll of two dice is the number of outcomes in C divided by 36.

1.1.1 Axioms of probability theory

Let $\emptyset$ be the **empty set**, that is, the event with no outcomes. We assume that the reader is familiar with the basic concepts of set theory such as **union** ($A \cup B$, the outcomes in either A or B) and **intersection** ($A \cap B$, the outcomes in both A and B).

Abstractly, a **probability** is a function that assigns numbers to events, which satisfies the following assumptions:

(i) For any event A, $0 \le P(A) \le 1$.
(ii) If Ω is the sample space then $P(\Omega) = 1$.

(iii) If A and B are **disjoint**, that is, the intersection $A \cap B = \emptyset$, then

$$P(A \cup B) = P(A) + P(B)$$

(iv) If $A_1, A_2, \ldots,$ is an infinite sequence of **pairwise disjoint events** (that is, $A_i \cap A_j = \emptyset$ when $i \neq j$) then

$$P\left(\bigcup_{i=1}^{\infty} A_i\right) = \sum_{i=1}^{\infty} P(A_i)$$

These assumptions are motivated by the **frequency interpretation of probability**, which states that if we repeat an experiment a large number of times then the fraction of times the event A occurs will be close to $P(A)$. To be precise, if we let $N(A, n)$ be the number of times A occurs in the first n trials then

$$P(A) = \lim_{n\to\infty} \frac{N(A, n)}{n} \tag{1.1}$$

In Chapter 6, we see that this result is a theorem called the law of large numbers. For the moment, we use this interpretation of $P(A)$ to explain the definition.

Given (1.1), assumptions (i) and (ii) are clear: the fraction of times a given event A occurs must be between 0 and 1, and if Ω has been defined properly (recall that it is the set of ALL possible outcomes) then the fraction of times something in Ω happens is 1. To explain (iii), note that if the events A and B are disjoint then

$$N(A \cup B, n) = N(A, n) + N(B, n)$$

since $A \cup B$ occurs if either A or B occurs but it is impossible for both to happen. Dividing by n and letting $n \to \infty$, we arrive at (iii).

Assumption (iii) implies that (iv) holds for a finite number of events, but for infinitely many events the last argument breaks down and this is a new assumption. Not everyone believes that Assumption (iv) should be used. However, without (iv) the theory of probability becomes much more difficult and less useful, so we impose this assumption and do not apologize further for it. In many cases the sample space is finite, so (iv) is not relevant anyway.

Example 1.3

Suppose we pick a letter at random from the word *TENNESSEE*. What is the sample space Ω and what probabilities should be assigned to the outcomes?

The sample space $\Omega = \{T, E, N, S\}$. To describe the probability it is enough to give the values for the individual outcomes since (iii) implies that $P(A)$ is the sum of the probabilities of the outcomes in A. Since there are nine letters in TENNESSEE, the probabilities are $P(\{T\}) = 1/9$, $P(\{E\}) = 4/9$, $P(\{N\}) = 2/9$, and $P(\{S\}) = 2/9$.

1.1.2 Basic properties of $P(A)$

Having introduced a number of definitions, we now derive some basic properties of probabilities and illustrate their use.

Property 1. *Let A^c be the* **complement** *of A, that is, the set of outcomes not in A. Then*

$$P(A) = 1 - P(A^c) \tag{1.2}$$

Proof. Let $A_1 = A$ and $A_2 = A^c$. Then $A_1 \cap A_2 = \emptyset$ and $A_1 \cup A_2 = \Omega$, so (iii) implies $P(A) + P(A^c) = P(\Omega) = 1$ by (ii). Subtracting $P(A)$ from each side of the equation gives the result. □

This formula is useful because sometimes it is easier to compute the probability of A^c. For an example, consider $A =$ "at least one 6." In this case $A^c =$ "no 6." There are $5 \cdot 5$ outcomes with no 6, so $P(A^c) = 25/36$ and $P(A) = 1 - 25/36 = 11/36$, as we computed before.

Property 2. *For any events A and B,*

$$P(A \cup B) = P(A) + P(B) - P(A \cap B) \tag{1.3}$$

Proof by picture:

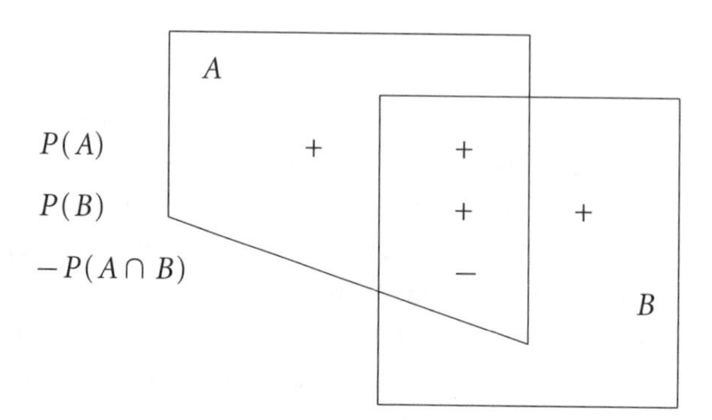

Intuitively, $P(A) + P(B)$ counts $A \cap B$ twice, so we have to subtract $P(A \cap B)$ to make the net number of times $A \cap B$ is counted equal to 1. □

Proof. To prove this result we note that by assumption (ii)

$$P(A) = P(A \cap B) + P(A \cap B^c)$$

$$P(B) = P(B \cap A) + P(B \cap A^c)$$

Adding the two equations and subtracting $P(A \cap B)$,

$$P(A) + P(B) - P(A \cap B) = P(A \cap B) + P(A \cap B^c) + P(B \cap A^c) = P(A \cup B)$$

which gives the desired equality. □

To illustrate Property 2, let $A =$ "the red die shows 6," and $B =$ "the green die shows 6." In this case $A \cup B =$ "at least one 6" and $A \cap B = \{(6, 6)\}$, so we have

$$P(A \cup B) = P(A) + P(B) - P(A \cap B) = \frac{1}{6} + \frac{1}{6} - \frac{1}{36} = \frac{11}{36}$$

The same principle applies to counting outcomes in events.

Example 1.4 A survey of 1,000 students revealed that 750 owned stereos, 450 owned cars, and 350 owned both. How many own either a car or a stereo?

Given a set A, we use $|A|$ to denote the number of points in A. The reasoning that led to (1.3) tells us that

$$|S \cup C| = |S| + |C| - |S \cap C| = 750 + 450 - 350 = 850$$

We can confirm this by drawing a picture:

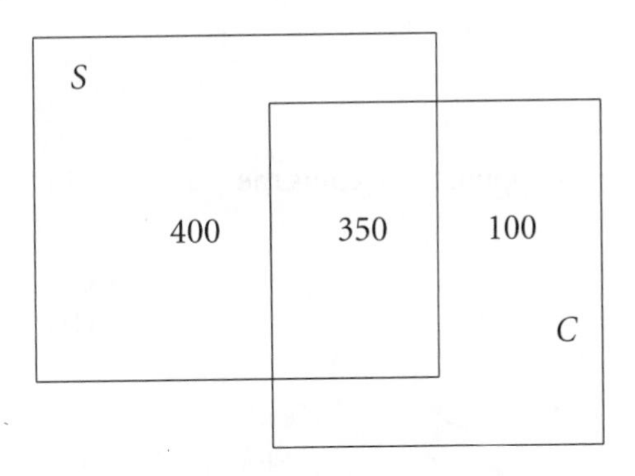

Property 3 (Monotonicity). *If $A \subset B$, that is, any outcome in A is also in B, then*

$$P(A) \leq P(B) \tag{1.4}$$

Proof. A and $A^c \cap B$ are disjoint, with union B, so assumption (iii) implies $P(B) = P(A) + P(A^c \cap B) \geq P(A)$ by (i). □

We write $A_n \uparrow A$ if $A_1 \subset A_2 \subset \cdots$ and $\cup_{i=1}^{\infty} A_i = A$. We write $A_n \downarrow A$ if $A_1 \supset A_2 \supset \cdots$ and $\cap_{i=1}^{\infty} A_i = A$.

Property 4 (Monotone limits). *If $A_n \uparrow A$ or $A_n \downarrow A$ then*

$$\lim_{n\to\infty} P(A_n) = P(A) \tag{1.5}$$

Proof. Let $B_1 = A_1$, and for $i \geq 2$ let $B_i = A_i \cap A_{i-1}^c$. The events B_i are disjoint, with $\cup_{i=1}^{\infty} B_i = A$, so (iv) implies

$$P(A) = \sum_{i=1}^{\infty} P(B_i) = \lim_{n\to\infty} \sum_{i=1}^{n} P(B_i) = \lim_{n\to\infty} P(A_n)$$

by (iii) since B_i, $1 \leq i \leq n$, are disjoint and their union is A_n.

To prove the second result, let $B_i = A_i^c$. We have $B_n \uparrow A^c$ so by (1.5) and (1.2), $\lim_{n\to\infty} P(B_n) = 1 - P(A)$. Since $P(B_n) = 1 - P(A_n)$, the desired result follows. □

1.2 Flipping coins and the World Series

Even simpler than rolling a die is flipping a coin, which produces one of two outcomes, called "heads" (H) or "tails" (T). If we flip two coins there are 4 outcomes:

	HH	HT TH	TT
Heads	2	1	0
Probability	1/4	1/2	1/4

Flipping three coins there are 8 possibilities:

	HHH	HHT HTH THH	TTH THT HTT	TTT
Heads	3	2	1	0
Probability	1/8	3/8	3/8	1/8

Our next problem concerns flipping four to seven coins.

Example 1.5

World Series. In this baseball event, the first team to win four games wins the championship. Obviously, the series may last 4, 5, 6, or 7 games. However, a fan who wants to buy a ticket would like to know what are the probabilities of each of these outcomes.

Here, we are assuming that the two teams are equally matched and ignoring potential complicating factors such as the advantage of playing at home or psychological factors that make the outcome of one game affect the next one. In short, we suppose that the games are decided by tossing a fair coin to determine whether team A or team B wins.

Four games. There are two possible ways this can happen: A wins all four games or B wins all four games. There are $2 \cdot 2 \cdot 2 \cdot 2 = 16$ possible outcomes and these are 2 of them, so $P(4) = 2/16 = 1/8$.

Five games. Here and in the next case we compute the probability that A wins in the specified number of games and then multiply by 2. There are 4 possible outcomes:

$$BAAAA, \quad ABAAA, \quad AABAA, \quad AAABA$$

$AAAAB$ is not possible since in that case the series would have ended in four games. There are $2^5 = 32$ outcomes, so $P(5) = 2 \cdot 4/32 = 1/4$.

Six games. In the next section we learn systematic ways of doing this, but for now we compute the probabilities by enumerating the possibilities:

$$\begin{array}{llll} BBAAAA & ABBAAA & AABBAA & AAABBA \\ BABAAA & ABABAA & AABABA & \\ BAABAA & ABAABA & & \\ BAAABA & & & \end{array}$$

The first column corresponds to outcomes in which B wins the first game, the second one to outcomes in which the first game B wins is the second game, etc. We then move the remaining win for B through its possibilities. There are 10 outcomes out of $2^6 = 64$ total, so remembering to multiply by 2 to account for the ways B can win in six games, $P(6) = 2 \cdot 10/64 = 5/16$.

Seven games. The analysis from the previous case becomes even messier here, so we instead observe that the probabilities for the four possible outcomes must add up to 1, so

$$P(7) = 1 - P(4) - P(5) - P(6) = 1 - \frac{2}{16} - \frac{4}{16} - \frac{5}{16} = \frac{5}{16}$$

As mentioned earlier, we are ignoring things that many fans think are important in determining the outcomes of the games, so our next step is to compare the probabilities just calculated with the observed frequencies of the duration of best-of-seven series in three sports. The numbers in parentheses give the number of series in our sample.

Games	4	5	6	7
Probabilities	0.125	0.25	0.3125	0.3125
World Series (94)	0.181	0.224	0.224	0.372
Stanley Cup (74)	0.270	0.216	0.230	0.284
NBA finals (57)	0.122	0.228	0.386	0.263

To determine whether or not the data agree with predictions, statisticians use a **chi-squared statistic:**

$$\chi^2 = \sum \frac{(o_i - e_i)^2}{e_i}$$

where o_i is the number of observations in category i and e_i is what the model predicts. The details of the test are beyond the scope of this book, so we just quote the results: the Stanley Cup data are very unusual (the probability of a chi-square score this large or larger has probability $p < 0.01$) due to the larger-than-expected number of four-game series. The World Series data do not fit the model well, but are not very unusual ($p > 0.05$). On the other hand, the NBA finals data look like what we expect to see. The excess of six-game series can be due just to chance.

Example 1.6

Birthday problem. There are 30 people at a party. Someone wants to bet you $10 that there are 2 people with exactly the same birthday. Should you take the bet?

To pose a mathematical problem we ignore February 29 which only comes in leap years, and suppose that each person at the party picks their birthday at random from the calendar. There are 365^{30} possible outcomes for this experiment. The number of outcomes in which all the birthdays are different is

$$365 \cdot 364 \cdot 363 \cdot \cdots \cdot 336$$

since the second person must avoid the first person's birthday, the third the first two birthdays, and so on, until the 30th person must avoid the 29 previous birthdays. Let D be the event that all birthdays are different. Dividing the number of outcomes in which all the birthdays are different by the total number of outcomes, we have

$$P(D) = \frac{365 \cdot 364 \cdot 363 \cdot \cdots \cdot 336}{365^{30}} = 0.293684$$

In words, only about 29% of the time all the birthdays are different, so you will lose the bet 71% of the time.

At first glance it is surprising that the probability of 2 people having the same birthday is so large, since there are only 30 people compared with 365 days on the calendar. Some of the surprise disappears if you realize that there are $(30 \cdot 29)/2 = 435$ pairs of people who are going to compare their birthdays. Let p_k be the probability that k people all have different birthdays. Clearly, $p_1 = 1$ and $p_{k+1} = p_k(365 - k)/365$. Using this recursion it is easy to generate the values of p_k for $1 \leq k \leq 40$.

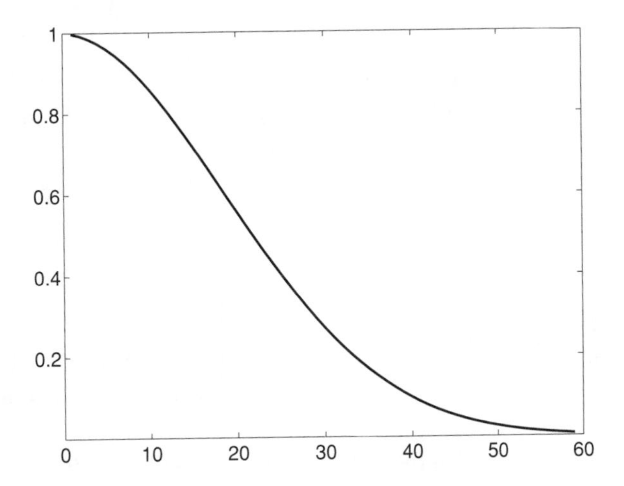

The graph shows the trends, but to get precise values a table is better:

1	1.00000	11	0.85886	21	0.55631	31	0.26955
2	0.99726	12	0.83298	22	0.52430	32	0.24665
3	0.99180	13	0.80559	23	0.49270	33	0.22503
4	0.98364	14	0.77690	24	0.46166	34	0.20468
5	0.97286	15	0.74710	25	0.43130	35	0.18562
6	0.95954	16	0.71640	26	0.40176	36	0.16782
7	0.94376	17	0.68499	27	0.37314	37	0.15127
8	0.92566	18	0.65309	28	0.34554	38	0.13593
9	0.90538	19	0.62088	29	0.31903	39	0.12178
10	0.88305	20	0.58856	30	0.29368	40	0.10877

1.3 Independence

Intuitively, two events A and B are independent if the occurrence of A has no influence on the probability of occurrence of B. The formal definition is A and B are **independent** if

$$P(A \cap B) = P(A)P(B)$$

To make the connection between the two definitions, we need to introduce the notion of conditional probability, which is discussed in more detail in Chapter 3.

Suppose we are told that the event A with $P(A) > 0$ occurs. Then the sample space is reduced from Ω to A and the probability that B will occur given that A has occurred is

$$P(B|A) = \frac{P(B \cap A)}{P(A)} \tag{1.6}$$

To explain this formula, note that (i) only the part of B that lies in A can possibly occur and (ii) since the sample space is now A, we have to divide by $P(A)$ to make $P(A|A) = 1$.

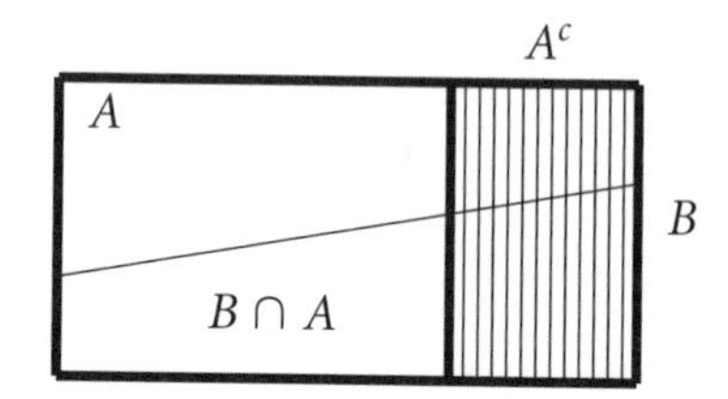

Suppose A and B are independent. In this case $P(A \cap B) = P(A)P(B)$, so

$$P(B|A) = \frac{P(A)P(B)}{P(A)} = P(B)$$

or in the words of the intuitive definition of independence, "the occurrence of A has no influence on the probability of the occurrence of B."

Turning to concrete examples, in each case it should be clear that the intuitive definition is satisfied, so we only check the formal one.

- Flip two coins. Let $A =$ "the first coin shows heads" and $B =$ "the second coin shows heads." $P(A) = 1/2$, $P(B) = 1/2$, $P(A \cap B) = 1/4$.
- Roll two dice. Let $A =$ "the first die shows 5" and $B =$ "the second die shows 2." $P(A) = 1/6$, $P(B) = 1/6$, $P(A \cap B) = 1/36$.
- Pick a card from a standard deck of 52 cards. Let $A =$ "the card is an ace" and $B =$ "the card is a spade" $P(A) = 1/13$, $P(B) = 1/4$, $P(A \cap B) = 1/52$.

Two examples of events that are not independent are

Example 1.7 **Draw two cards from a deck.** Let $A =$ "the first card is a spade" and $B =$ "the second card is a spade." Then $P(A) = 1/4$ and $P(B) = 1/4$, but

$$P(A \cap B) = \frac{13 \cdot 12}{52 \cdot 51} < \left(\frac{1}{4}\right)^2$$

To derive the answer, we note that there are $52 \cdot 51$ outcomes for drawing two cards from the deck when we keep track of the order they are drawn, while there are $13 \cdot 12$ outcomes that result in two spades. A second approach is to note that we have a probability of $P(A) = 13/52$ of getting a spade the first time and if we succeed then the probability on the second draw is $P(B|A) = 12/51$. Rearranging the definition of conditional probability in (1.6)

$$P(A \cap B) = P(A)P(B|A)$$

Thus, these two events are not independent, since getting a spade the first time reduces the fraction of spades in the deck and makes it harder to get a spade the second time.

Example 1.8

Roll two dice. Let $A =$ "the sum of the two dice is 9" and $B =$ "the first die is 2." $A = \{(6, 3), (5, 4), (4, 5), (3, 6)\}$, so $P(A) = 4/36$. $P(B) = 1/6$, but $P(A \cap B) = 0$ since $(2, 7)$ is impossible.

In general, if A and B are disjoint events that have positive probabilities then they are not independent since $P(A)P(B) > 0 = P(A \cap B)$.

There are two ways of extending the definition of independence to more than two events. $A_1, \ldots, A_n$ are said to be **pairwise independent** if for each $i \neq j$, $P(A_i \cap A_j) = P(A_i)P(A_j)$; that is, each pair is independent. $A_1, \ldots, A_n$ are said to be **independent** if for any $1 \leq i_1 < i_2 < \cdots < i_k \leq n$, we have

$$P(A_{i_1} \cap \cdots \cap A_{i_k}) = P(A_{i_1}) \cdots P(A_{i_k})$$

If we flip n coins and let $A_i =$ "the ith coin shows heads," then the events A_i are independent since $P(A_i) = 1/2$ and $P(A_{i_1} \cap \cdots \cap A_{i_k}) = 1/2^k$. We have already seen an example of events that are pairwise independent but not independent.

Example 1.9

Birthdays. Let $A =$ "Alice and Betty have the same birthday," $B =$ "Betty and Carol have the same birthday," and $C =$ "Carol and Alice have the same birthday." Each pair of events is independent but the three are not.

Since there are 365 ways 2 girls can have the same birthday out of 365^2 possibilities (as in Example 1.6, we are assuming that leap year does not exist and that all the birthdays are equally likely), $P(A) = P(B) = P(C) = 1/365$. Likewise, there are 365 ways all 3 girls can have the same birthday out of 365^3 possibilities, so

$$P(A \cap B) = \frac{1}{365^2} = P(A)P(B)$$

that is, A and B are independent. Similarly, B and C are independent and C and A are independent, so A, B, and C are pairwise independent. The three events

A, B, and C are not independent; however, since $A \cap B = A \cap B \cap C$ and hence

$$P(A \cap B \cap C) = \frac{1}{365^2} \neq \left(\frac{1}{365}\right)^3 = P(A)P(B)P(C)$$

Example 1.10

Roll three dice. Let $A =$ "the numbers on the first and second add to 7," $B =$ "the numbers on the second and third add to 7," and $C =$ "the numbers on the third and first add to 7." Again each pair of events is independent but the three are not.

To check that A and B are independent, note that no matter what the values i and j are on the first two dice there is probability $1/6$ that the third die is $7 - j$. Similar arguments show that B and C are independent and A and C are independent. To show that the three events are not independent, we note that $A \cap B \cap C = \emptyset$. Let i, j, k be the values for the three dice. We claim that if A and B occur then $i + k$ is even, so C is impossible. To check this, note that if j is odd then i and k are even, while if j is even then i and k are odd.

The last two examples are somewhat unusual. However, the moral of the story is that to show several events are independent, you have to check more than just that each pair is independent.

1.4 Random variables and distributions

A **random variable** is a numerical value determined by the outcome of an experiment. Consider the following four examples:

- Roll two dice and let $X =$ the sum of the two numbers that appear.
- Roll a die until a 4 appears and let $X =$ the number of rolls we need.
- Flip a coin 10 times and let $X =$ the number of heads we get.
- Draw 13 cards out of a deck and let $X =$ the number of hearts we get.

In these cases X is a **discrete random variable**. That is, there is a finite or countable sequence of possible values. In contrast, the height of a randomly chosen person or the time a person spends waiting for the bus in the morning are **continuous random variables**, since the value could be any positive real number. In this book we are primarily concerned with discrete random variables. Continuous random variables do not make an appearance until Chapter 5.

The **distribution** of a discrete random variable is described by giving the value of $P(X = x)$ for all values of x. In each case, we only give the values of $P(X = x)$ when $P(X = x) > 0$. The other values that we do not mention are 0. We begin with the first two examples.

Example 1.11 **Roll two dice.** Let $X =$ the sum of the two numbers that appear.

$$\begin{array}{cccccc}
(1,1) & (2,1) & (3,1) & \mathbf{(4,1)} & (5,1) & (6,1) \\
(1,2) & (2,2) & \mathbf{(3,2)} & (4,2) & (5,2) & (6,2) \\
(1,3) & \mathbf{(2,3)} & (3,3) & (4,3) & (5,3) & (6,3) \\
\mathbf{(1,4)} & (2,4) & (3,4) & (4,4) & (5,4) & (6,4) \\
(1,5) & (2,5) & (3,5) & (4,5) & (5,5) & (6,5) \\
(1,6) & (2,6) & (3,6) & (4,6) & (5,6) & (6,6)
\end{array}$$

Using the table of outcomes, it is easy to see

x	2	3	4	5	6	7	8	9	10	11	12
$P(X = x)$	$\frac{1}{36}$	$\frac{2}{36}$	$\frac{3}{36}$	$\frac{4}{36}$	$\frac{5}{36}$	$\frac{6}{36}$	$\frac{5}{36}$	$\frac{4}{36}$	$\frac{3}{36}$	$\frac{2}{36}$	$\frac{1}{36}$

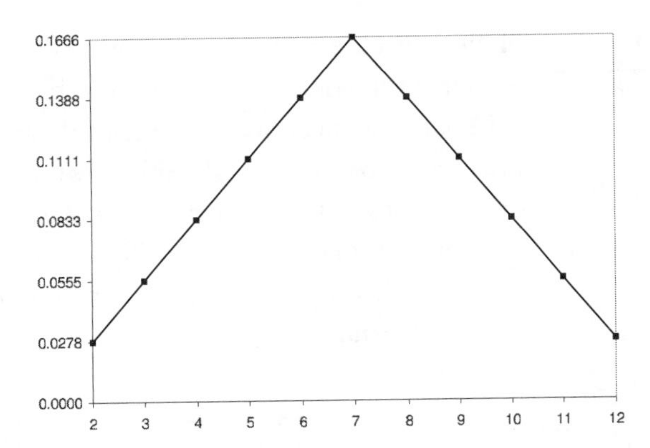

To check this, note that the outcomes with a given sum are diagonal lines in the square. For example, the four outcomes in boldface are the ones for which the sum is 5.

Example 1.12 **Geometric distribution.** If we repeat an experiment with probability p of success until a success occurs, the number of trials required, N, has

$$P(N = n) = (1 - p)^{n-1} p \quad \text{for } n = 1, 2, \ldots$$

In words, N has a geometric distribution with parameter p, a phrase we abbreviate as $N =$ geometric(p).

To check the formula, note that in order to first have success on trial n, we must have $n - 1$ failures followed by a success, which has probability $(1 - p)^{n-1} p$. In the example at the beginning of the section, success is rolling a 4, so $p = 1/6$. The next graph gives a picture of the distribution in this case.

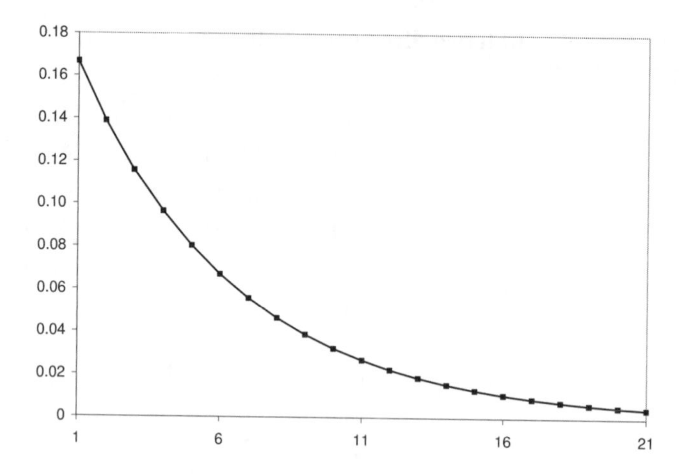

For an example of the use of the geometric distribution, we consider

Example 1.13

Birthday problem, II. How large must the group be so that there is a probability >0.5 that someone will have the same birthday as you do?

In our first encounter with the birthday problem it was surprising that the size needed to have 2 people with the same birthday was so small. This time the surprise goes in the other direction. Assuming 365 equally likely birthdays, a naive guess is that 183 people will be enough. However, in a group of n people the probability that all will fail to have your birthday is $(364/365)^n$. Setting this equal to 0.5 and solving,

$$n = \frac{\log(0.5)}{\log(364/365)} = \frac{-0.69314}{-0.0027435} = 252.7$$

So we need 253 people. The "problem" is that many people in the group will have the same birthday, so the number of different birthdays is smaller than the size of the group.

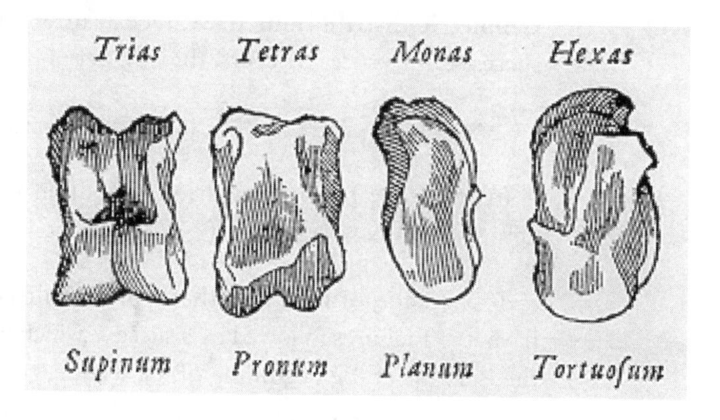

Example 1.14

Astragali. Board games involving chance were known in Egypt, 3000 B.C. The element of chance needed for these games was at first provided by tossing astragali, the ankle bones of sheep. These bones could come to rest on only four sides, the other two sides being rounded. The upper side of the bone, broad and slightly convex, counted four (tetras); the opposite side, broad and slightly concave, counted three (trias); the lateral side, flat and narrow, one (monas), and the opposite narrow lateral side, which is slightly hollow, six (hexas).

The outcomes of this experiment are $\Omega = \{1, 3, 4, 6\}$. There is no reason to suppose that all four sides have the same probability, so our model will have probabilities for the four outcomes $p_1, p_3, p_4, p_6 \geq 0$ that have $p_1 + p_3 + p_4 + p_6 = 1$. To define the probability of an event A, we let

$$P(A) = \sum_{i \in A} p_i$$

In words, we add up the probabilities of the outcomes in A. With a little thought we see that any probability with a finite set of outcomes has this form.

Example 1.15

English letter frequencies. In text written in English the 26 letters in the alphabet occur with the following frequencies:

E	13.0%	H	3.5%	W	1.6%
T	9.3%	L	3.5%	V	1.3%
N	7.8%	C	3.0%	B	0.9%
R	7.7%	F	2.8%	X	0.5%
O	7.4%	P	2.7%	K	0.3%
I	7.4%	U	2.7%	Q	0.3%
A	7.3%	M	2.5%	J	0.2%
S	6.3%	Y	1.9%	Z	0.1%
D	4.4 %	G	1.6%		

From this it follows that vowels (A, E, I, O, U) are used $7.3 + 13.0 + 7.4 + 7.4 + 2.7 = 37.8\%$ of the time.

Example 1.16

Scrabble. In this game, there are 100 tiles with the following distribution. The first number after the letter is its point value and the second is the number of tiles.

E	1	12	U	1	4	V	4	2
A	1	9	D	2	4	W	4	2
I	1	9	G	2	3	Y	4	2
O	1	8	B	3	2	K	5	1
N	1	6	C	3	2	J	8	1
R	1	6	M	3	2	X	8	1
T	1	6	P	3	2	Q	10	1
L	1	4	F	4	2	Z	10	1
S	1	4	H	4	2	blank	0	2

In Scrabble, vowels are $12 + 9 + 9 + 8 + 4 = 42\%$ of the letters. The number of points on a randomly chosen tile has the following distribution:

0	1	2	3	4	5	8	10
0.02	0.68	0.07	0.08	0.10	0.01	0.02	0.02

Example 1.17

Benford's law is named after the late Dr. Frank Benford, a physicist at General Electric Company. In 1938, he noticed that pages of logarithms corresponding to numbers starting with 1 were dirtier and more worn than other pages. He examined 20,229 data sets and developed a prediction about the observed distribution of first digits:

$$p_k = \log_{10}(k + 1) - \log_{10}(k) \quad \text{for } k = 1, 2, \ldots, 9$$

This is a probability distribution because $p_k \geq 0$ and $\sum_{k=1}^{9} p_k = \log_{10}(10) - \log_{10}(1)$. The numerical values of the probabilities are

1	2	3	4	5	6	7	8	9
0.3010	0.1761	0.1249	0.0969	0.0792	0.0669	0.0580	0.0512	0.0458

Some of the many examples that are supposed to follow Benford's law are census populations of 3,259 counties, 308 numbers from *Reader's Digest*, areas of 335 rivers, 342 addresses of *American Men of Science*. The next table compares the percentages of the observations in the first five categories to Benford's law:

	1	2	3	4	5
Census	33.9	20.4	14.2	8.1	7.2
Reader's Digest	33.4	18.5	12.4	7.5	7.1
Rivers	31.0	16.4	10.7	11.3	7.2
Benford's law	30.1	17.6	12.5	9.7	7.9
Addresses	28.9	19.2	12.6	8.8	8.5

The fits are far from perfect, but in each case Benford's law matches the general shape of the observed distribution.

Why should Benford's law hold? It is a mathematical fact that the first digits of 2^n follow Benford's law. This can be established by showing that the fractional parts of $\log_{10}(2^m)$ for $1 \le m \le n$ (that is, the part of the number to the right of the decimal point) are, for large n, approximately uniformly distributed on $(0, 1)$.

There are two good general explanations:

(i) If the Dow Jones average or some other statistic grows at a rate r, that is, $f(t) = C e^{rt}$, then the amount of time it takes for the first digit to change from k to $k+1$ is $[\ln(k+1) - \ln(k)]/r$. When we divide by the total time to go from 1 to 10, the result is Benford's law.

(ii) Ted Hill proved (see *Statistical Science* 10, 354–363) that if the first-digit distribution does not depend on the units used then Benford's law is the only possibility.

1.5 Expected value

The **expected value** of X, or **mean** of X, is defined to be

$$EX = \sum_x x P(X = x) \tag{1.7}$$

In words, we multiply each possible value by its probability and sum.

Example 1.18 **Roulette.** If you play roulette and bet \$1 on black then you win \$1 with probability 18/38 and you lose \$1 with probability 20/38, so the expected value of your winnings X is

$$EX = 1 \cdot \frac{18}{38} + (-1) \cdot \frac{20}{38} = \frac{-2}{38} = -0.0526$$

The expected value has a meaning much like the frequency interpretation of probability: in the long run you will lose about 5.26 cents per play.

To make a precise statement we need a definition. Suppose $X_1, \ldots, X_n$ are independent and have the same distribution as X; that is,

$$P(X_1 = x_1, \ldots, X_n = x_n) = P(X = x_1) \cdots P(X = x_n)$$

The law of large numbers, which is proved in Chapter 6, says that when n is large the average of the values we have observed, $(X_1 + \cdots + X_n)/n$, will be close to EX with high probability. In the roulette example, if we let X_i be your winnings on the ith play then this law says that $(X_1 + \cdots + X_n)/n$ will be close to -0.0526.

Example 1.19

Roll one die. Let X be the number that appears on the die. $P(X = x) = 1/6$ for $x = 1, 2, 3, 4, 5, 6$, so

$$EX = 1 \cdot \frac{1}{6} + 2 \cdot \frac{1}{6} + 3 \cdot \frac{1}{6} + 4 \cdot \frac{1}{6} + 5 \cdot \frac{1}{6} + 6 \cdot \frac{1}{6} = \frac{21}{6} = 3\frac{1}{2}$$

In this case the expected value is just the average of the six possible values.

To deal with more than one die we use the following fact, which is proved in Chapter 6.

Theorem 1.1. *If $X_1, \ldots, X_n$ are random variables then*

$$E(X_1 + \cdots + X_n) = EX_1 + \cdots + EX_n \tag{1.8}$$

From this and Example 1.19, it follows that if we roll two dice the expected value of the sum is 7. As another example, let $X_i = 1$ if the ith flip of a coin is heads and 0 otherwise. $X_1 + \cdots + X_n$ is the number of heads in n tosses. Since $EX_i = 1/2$ for a fair coin, the last result implies that

$$E(X_1 + \cdots + X_n) = nEX_i = \frac{n}{2}$$

Example 1.20

Scrabble. As we computed in Example 1.16, the point value of a randomly chosen Scrabble tile has the following distribution:

Value	0	1	2	3	4	5	8	10
Probability	0.02	0.68	0.07	0.08	0.10	0.01	0.02	0.02

The expected value is

$$= 0.68 + 0.14 + 0.24 + 0.4 + 0.05 + 0.16 + 0.2 = 1.87$$

This means that when we draw 7 letters to start the game the average numbers of points on our rack will be $7(1.87) = 13.09$. Of course, on any play of the game the number of points may be more or less than 13. The law of large numbers implies that if we keep records for a large number of games then the average we see on our first draws will be close to the expected value.

Example 1.21

Fair division of a bet on an interrupted game. Pascal and Fermat were sitting in a café in Paris playing the simplest of all games, flipping a coin. Each had put up a bet of 50 francs and the first to get 10 points wins. Fermat was winning 8 points to 7 when he received an urgent message that a friend was sick and he must rush to his hometown of Toulouse. The carriage man who delivered the message offered to take him, but only if he would leave immediately. Later in correspondence between the two men, the problem arose: How should the money bet (100 francs) be divided?

Fermat came up with the reasonable idea that the fraction of the stakes that each receives should be equal to the probability it would have won the match. In the case under consideration, it is easier to calculate the probability that Pascal (P) wins. In order for Pascal to win by a score of 10–8, he must win three flips in a row: $P P P$, an event of probability 1/8. To win 10–9, he can lose one flip but not the last one: $F P P P$, $P F P P$, $P P F P$, which has probability $3/16$. Adding the two we see that Pascal wins with probability $5/16$ and should get that fraction of the total bet, that is, $(5/16)(100) = 31.25$.

Example 1.22 ***Deal or No Deal.*** In this TV game show there are 26 briefcases with amounts of money indicated in the first, third, and fifth columns of the next table. You pick one briefcase and then pick five others to open. At that point they offer you an amount of money. If you take it the game ends. If not, then you open more briefcases. The numbers in the second, fourth, and sixth columns are the rounds on which I opened those briefcases when I played the game online at nbc.com:

0.01	2	300	3	75,000	4
1	1	400	4	100,000	2
5	2	500	3	200,000	
10		750	3	300,000	2
25	2	1,000	9	400,000	5
50	4	5,000	1	500,000	3
75	5	10,000	1	750,000	7
100	8	25,000	6	1,000,000	1
200	1	50,000	1		

The expected value of the money in the briefcase you pick is 131,477. The next table gives the offers I got from the online game compared with the expected value after each of the rounds.

1	25,866	117,660
2	35,158	122,074
3	30,492	120,970
4	46,446	152,910
5	48,806	162,685
6	64,675	190,222
7	21,620	50,277
8	40,872	67,003
9	62,003	100,005

Notice that in all cases the offer is considerably less than the expected value. After the ninth round there are two briefcases left, one with 200,000 and other with 10. If I were playing for real I might have taken the 62,003 offer for sure but I stuck with the higher expected value and won 200,000.

Example 1.23

Geometric distribution. When $P(N = n) = (1 - p)^{n-1} p$, for $n = 1, 2, 3, \ldots$, we have $EN = 1/p$.

This answer is intuitive. We have a probability p of success on each trial, so in n trials we have an average of np successes and if we want $np = 1$, we need $n = 1/p$. To get this from the definition, we begin with the sum of the geometric series

$$\sum_{k=0}^{\infty} x^k = \frac{1}{1-x}$$

and differentiate with respect to x to get

$$\sum_{k=0}^{\infty} kx^{k-1} = \frac{1}{(1-x)^2}$$

Dropping the $k = 0$ term from the left since it is 0 and setting $x = 1 - p$,

$$\sum_{k=1}^{\infty} k(1-p)^{k-1} = \frac{1}{p^2}$$

Multiplying each side by p we have

$$\sum_{k=1}^{\infty} kP(N = k) = \frac{1}{p}$$

Example 1.24

China's one-child policy. In order to limit the growth of its population, the Chinese government decided to limit families to having just one child. An alternative that was suggested was the "one-son" policy: as long as a woman has only female children she is allowed to have more children. One concern voiced about this policy was that no family would have more than one son, but many families would have several girls. This concern leads to our question: How would the one-son policy affect the ratio of male to female births?

To simplify the problem we assume that a family will keep having children until it has a male child. Assuming that male and female children are equally likely and the sexes of successive children are independent, the total number of children has a geometric distribution with success probability $p = 1/2$, so by the previous example the expected number of children is $1/p = 2$. There is always one male child, so the expected number of female children is $2 - 1 = 1$.

Does this continue to hold if some families stop before they have a male child? Consider for simplicity the case in which a family will stop when they have a male child or a total of three children. There are 4 outcomes:

$$P(M) = \frac{1}{2}$$

$$P(FM) = \frac{1}{4}$$

$$P(FFM) = \frac{1}{8}$$

$$P(FFF) = \frac{1}{8}$$

The average number of male children is $1/2 + 1/4 + 1/8 = 7/8$, while the average number of female children is $1(1/4) + 2(1/8) + 3(1/8) = 7/8$.

The last calculation makes the equality of the expected values look like a miracle, but it is not, and the claim holds true if a family with k female children continues with probability p_k and stops with probability $1 - p_k$. To explain this intuitively, if we replace M by $+1$ and F by -1, then childbirth is a fair game. For the stopping rules under consideration the average winnings when we stop have mean 0; that is, the expected number of male children equals the expected number of female children.

1.6 Moments and variance

In this section we are interested in the expected values of various functions of random variables. The most important of these are the variance and the standard deviation, which give an idea about how spread out the distribution is. The first basic fact we need in order to do computations is

Theorem 1.2. *If X has a discrete distribution and $Y = r(X)$ then*

$$EY = \sum_x r(x)P(X = x) \tag{1.9}$$

Proof. $P(Y = y) = \sum_{x:r(x)=y} P(X = x)$. Multiplying both sides by y and summing gives

$$\begin{aligned} EY &= \sum_y y\,P(Y = y) = \sum_y y \sum_{x:r(x)=y} P(X = x) \\ &= \sum_y \sum_{x:r(x)=y} r(x)P(X = x) = \sum_x r(x)P(X = x) \end{aligned}$$

since the double sum is just a clumsy way of summing over all the possible values of x. □

If $r(x) = x^k$, $E(X^k)$ is the k**th moment of** X. When $k = 1$, this is the first moment or mean of X.

Example 1.25 **Roll one die.** Let X be the resulting number. Compute EX^2.

$$EX^2 = \frac{1}{6}(1 + 4 + 9 + 16 + 25 + 36) = \frac{91}{6} = 15.1666$$

To prepare for our next topic we need the following properties:

$$E(X + b) = EX + b \qquad E(aX) = aEX \tag{1.10}$$

In words, if we add 5 to a random variable then we add 5 to its expected value. If we multiply a random variable by 3 we multiply its expected value by 3.

Proof. For the first one, we note that

$$E(X + b) = \sum_x (x + b)\, P(X = x)\, dx$$
$$= \sum_x xP(X = x) + \sum_x bP(X = x) = EX + b$$

The second is easier:

$$E(aX) = a \sum_x xP(X = x) = aEX$$ □

If $EX^2 < \infty$ then the **variance** of X is defined to be

$$\text{var}(X) = E(X - EX)^2$$

To illustrate this concept, we consider some examples. But first, we need a formula that enables us to compute the variance more easily.

$$\text{var}(X) = EX^2 - (EX)^2 \tag{1.11}$$

Proof. Letting $\mu = EX$ to make the computations easier to see, we have

$$\text{var}(X) = E(X - \mu)^2 = E(X^2 - 2\mu X + \mu^2) = EX^2 - 2\mu EX + \mu^2$$

by (1.10) and the facts that $E(-2\mu X) = -2\mu EX$ and $E(\mu^2) = \mu^2$. Substituting $\mu = EX$ now gives the result. □

The reader should note that EX^2 means the expected value of X^2 and in the proof $E(X - \mu)^2$ means the expected value of $(X - \mu)^2$. When we want the square of the expected value we write $(EX)^2$. This convention is designed to cut down on parentheses.

The variance measures how spread out the distribution of X is. To begin to explain this statement, we show that

$$\text{var}(X+b) = \text{var}(X) \qquad \text{var}(aX) = a^2\text{var}(X) \tag{1.12}$$

In words, the variance is not changed by adding a constant to X, but multiplying X by a multiplies the variance by a^2.

Proof. If $Y = X + b$ then the mean of Y, $\mu_Y = \mu_X + b$ by (1.10), so

$$\text{var}(X+b) = E[(X+b) - (\mu_X + b)]^2 = E(X-\mu_X)^2 = \text{var}(X)$$

If $Y = aX$ then $\mu_Y = a\mu_X$ by (1.10), so

$$\text{var}(aX) = E[(aX - a\mu_X)^2] = a^2 E(X-\mu_X)^2 = a^2\text{var}(X) \qquad \square$$

The scaling relationship (1.12) shows that if X is measured in feet then the variance is measured in square feet. This motivates the definition of the **standard deviation** $\sigma(X) = \sqrt{\text{var}(X)}$, which is measured in the same units as X and has a nicer scaling property:

$$\sigma(aX) = |a|\sigma(X) \tag{1.13}$$

We get the absolute value here since $\sqrt{a^2} = |a|$.

Example 1.26

Roll one die. Let X be the resulting number. Find the variance and standard deviation of X.

Examples 1.19 and 1.25 tell us that $EX = 7/2$ and $EX^2 = 91/6$, so

$$\text{var}(X) = EX^2 - (EX)^2 = \frac{91}{6} - \frac{49}{4} = \frac{105}{36} = 2.9166$$

and $\sigma(X) = \sqrt{\text{var}(X)} = 1.7078$. The standard deviation $\sigma(X)$ gives the size of the "typical deviation from the mean." To explain this, we note that the deviation from the mean

$$|X-\mu| = \begin{cases} 0.5 & \text{when } X = 3, 4 \\ 1.5 & \text{when } X = 2, 5 \\ 2.5 & \text{when } X = 1, 6 \end{cases}$$

so $E|X-\mu| = 1.5$. The standard deviation $\sigma(X) = \sqrt{E|X-\mu|^2}$ is a slightly less intuitive way of averaging the deviations $|X-\mu|$ but, as we see later, is one that has nicer properties.

Example 1.27

Scrabble. As we computed in Example 1.16, the point value of a randomly chosen Scrabble tile has the following distribution:

Value	0	1	2	3	4	5	8	10
Probability	0.02	0.68	0.07	0.08	0.10	0.01	0.02	0.02

In Example 1.20, we have computed that $EX = 1.87$.

$$\begin{aligned} EX^2 &= 0.68 + 4(0.07) + 9(0.08) + 16(0.10) + 25(0.01) \\ &\quad + 64(0.02) + 100(0.02) = 9.06 \end{aligned}$$

so the variance is $9.06 - (1.87)^2 = 5.5631$ and the standard deviation is $\sqrt{5.5631} = 2.35$.

Example 1.28

Geometric distribution. Suppose $N =$ geometric(p). That is, $P(N = n) = (1-p)^{n-1}p$ for $n = 1, 2, \ldots,$ and 0 otherwise. Compute the variance and standard deviation of N.

To compute the variance, we begin, as in Example 1.23, by observing that

$$\sum_{n=0}^{\infty} x^n = (1-x)^{-1}$$

Differentiating this identity twice and noticing that the $n = 0$ term in the first derivative is 0 gives

$$\sum_{n=1}^{\infty} nx^{n-1} = (1-x)^{-2} \qquad \sum_{n=1}^{\infty} n(n-1)x^{n-2} = 2(1-x)^{-3}$$

Setting $x = 1 - p$ gives

$$\sum_{n=1}^{\infty} n(1-p)^{n-1} = p^{-2} \qquad \sum_{n=1}^{\infty} n(n-1)(1-p)^{n-2} = 2p^{-3}$$

Multiplying both sides by p in the first case and $p(1-p)$ in the second, we have

$$EN = \sum_{n=1}^{\infty} n(1-p)^{n-1}p = p^{-1}$$

$$E[N(N-1)] = \sum_{n=1}^{\infty} n(n-1)(1-p)^{n-1}p = 2p^{-2}(1-p)$$

From this it follows that

$$E N^2 = E[N(N-1)] + E N = \frac{2-2p}{p^2} + \frac{1}{p} = \frac{2-p}{p^2}$$

$$\text{var}(N) = E N^2 - (E N)^2 = \frac{2-p}{p^2} - \frac{1}{p^2} = \frac{1-p}{p^2}$$

Taking the square root, we see that $\sigma(X) = \sqrt{1-p}/p$.

Example 1.29 **New York Yankees 2004 salaries.** Salaries are in units of M (millions of dollars per year) and, for convenience, have been truncated at the thousands place.

A. Rodriguez	21.726	D. Jeter	18.6
M. Mussina	16	K. Brown	15.714
J. Giambi	12.428	B. Williams	12.357
G. Sheffield	12.029	M. Rivera	10.89
J. Posada	9	J. Vazquez	9
J. Contreras	9	J. Olerud	7.7
H. Matsui	7	S. Karsay	6
E. Loazia	4	T. Gordon	3.5
P. Quantrill	3	K. Lofton	2.985
J. Lieber	2.7	T. Lee	2
G. White	1.925	F. Heredia	1.8
R. Sierra	1	M. Cairo	0.9
J. Falherty	0.775	T. Clark	0.75
E. Wilson	0.7	O. Hernandez	0.5
D. Osborne	0.45	C.J. Nitowski	0.35
J. DePaula	0.302	B. Crosby	0.301

The total team salary is 183.355 M. Dividing by 32 players gives a mean of 6.149 M. The second moment is 73.778 M^2, so the variance is $73.778 - (6.149)^2 = 35.961$ M^2 and the standard deviation is 5.996 M. The next graph shows the fraction of Yankees with salaries $\geq x$ M.

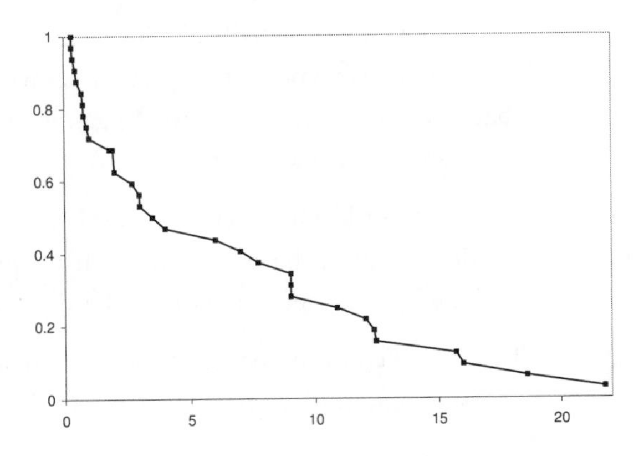

1.7 Exercises

Basic definitions

1. A man receives presents from his three children, Allison, Betty, and Chelsea. To avoid disputes he opens the presents in a random order. What are the possible outcomes?

2. Suppose we pick a number at random from the phone book and look at the last digit. (a) What is the set of outcomes and what probability should be assigned to each outcome? (b) Would this model be appropriate if we were looking at the first digit?

3. Two students arrive late for a math final exam with the excuse that their car had a flat tire. Suspicious, the professor says "each one of you write down on a piece of paper which tire was flat." What is the probability that both students pick the same tire?

4. Suppose we roll a red die and a green die. What is the probability that the number on the red die is larger than the number on the green die?

5. Two dice are rolled. What is the probability that (a) the two numbers will differ by 1 or less and (b) the maximum of the two numbers will be 5 or larger?

6. If we flip a coin 5 times, what is the probability that the number of heads is an even number (that is, divisible by 2)?

7. The 1987 World Series was tied at two games a piece before the St. Louis Cardinals won the fifth game. According to the Associated Press, "The numbers of history support the Cardinals and the momentum they carry. Whenever the series has been tied 2–2 the team that won the fifth game won the series 71% of the time." If momentum is not a factor and each team has a 50% chance of winning each game, what is the probability that the Game 5 winner will win the series?

8. Two boys are repeatedly playing a game that they each have probability 1/2 of winning. The first person to win 5 games wins the match. What is the probability that Al will win if (a) he has won 4 games and Bobby has won 3 and (b) he leads by a score of 3 games to 2?

9. 20 families live in a neighborhood: 4 have 1 child, 8 have 2 children, 5 have 3 children, and 3 have 4 children. If we pick a child at random, what is the probability that they come from a family with 1, 2, 3, 4 children?

10. In Galileo's time people thought that when three dice were rolled, a sum of 9 and a sum of 10 had the same probability since each could be obtained

in 6 ways:

$$9: \quad 1+2+6,\ 1+3+5,\ 1+4+4,\ 2+2+5,\ 2+3+4,\ 3+3+3$$

$$10: \quad 1+3+6,\ 1+4+5,\ 2+4+4,\ 2+3+5,\ 2+4+4,\ 3+3+4$$

Compute the probabilities of these sums and show that 10 is a more likely total than 9.

11. Suppose we roll three dice. Compute the probability that the sum is (a) 3, (b) 4, (c) 5, (d) 6, (e) 7, (f) 8, (g) 9, (h) 10.

12. In a group of students, 25% smoke cigarettes, 60% drink alcohol, and 15% do both. What fraction of students have at least one of these bad habits?

13. In a group of 320 high school graduates, only 160 went to college but 100 of the 170 men did. How many women did not go to college?

14. In the freshman class, 62% of the students take math, 49% take science, and 38% take both science and math. What percentage takes at least one science or math course?

15. 24% of people have American Express cards, 61% have Visa cards, and 8% have both. What percentage of people have at least one credit card?

16. Suppose $\Omega = \{a, b, c\}$, $P(\{a, b\}) = 0.7$, and $P(\{b, c\}) = 0.6$. Compute the probabilities of $\{a\}$, $\{b\}$, and $\{c\}$.

17. Suppose A and B are disjoint with $P(A) = 0.3$ and $P(B) = 0.5$. What is $P(A^c \cap B^c)$?

18. Given two events A and B with $P(A) = 0.4$ and $P(B) = 0.7$, what are the maximum and minimum possible values for $P(A \cap B)$?

Independence

19. Suppose we draw 2 cards out of a deck of 52. Let $A =$ "the first card is an ace" and $B =$ "the second card is a spade." Are A and B independent?

20. A family has 3 children, each of whom is a boy or a girl with probability 1/2. Let $A =$ "there is at most 1 girl" $B =$ "the family has children of both sexes." (a) Are A and B independent? (b) Are A and B independent if the family has 4 children?

21. Suppose we roll a red and a green die. Let $A =$ "the red die shows a 2 or a 5" and $B =$ "the sum of the two dice is at least 7." Are A and B independent?

22. Roll two dice. Let $A =$ "the sum is even" and $B =$ "the sum is divisible by 3," that is, $B = \{3, 6, 9, 12\}$. Are A and B independent?

23. Roll two dice. Let $A =$ "the first die is odd," $B =$ "the second die is odd," and $C =$ "the sum is odd." Show that these events are pairwise independent but not independent.

24. Nine children are seated at random in three rows of three desks. Let $A =$ "Al and Bobby sit in the same row" and $B =$ "Al and Bobby both sit at one of the four corner desks." Are A and B independent?

25. Two students, Alice and Betty, are registered for a statistics class. Alice attends 80% of the time, Betty 60% of the time, and their absences are independent. On a given day, what is the probability that (a) at least one of these students is in class and (b) exactly one of them is there?

26. Let A and B be two independent events with $P(A) = 0.4$ and $P(A \cup B) = 0.64$. What is $P(B)$?

27. Three students each have probability 1/3 of solving a problem. What is the probability that at least one of them will solve the problem?

28. Three independent events have probabilities 1/4, 1/3, and 1/2. What is the probability that exactly one will occur?

29. Three missiles are fired at a target. They will hit it with probabilities 0.2, 0.4, and 0.6. Find the probability that the target is hit by (a) three, (b) two, (c) one, and (d) no missiles.

30. Three couples that were invited to dinner will independently show up with probabilities 0.9, 0.89, and 0.75. Let N be the number of couples that show up. Calculate the probability $P(N)$ with $N = 3, 2, 1, 0$.

31. A college student takes 4 courses a semester for 8 semesters. In each course she has a probability 1/2 of getting an A. Assuming her grades in different courses are independent, what is the probability that she will have at least one semester with all A's?

32. When Al and Bob play tennis, Al wins a set with probability 0.7 while Bob wins with probability 0.3. What is the probability that Al will be the first to win (a) two sets and (b) three sets?

33. Chevalier de Méré made money betting that he could "roll at least one 6 in 4 tries." When people got tired of this wager he changed it to "roll at least one double 6 in 24 tries," but then he started losing money. Compute the probabilities of winning these two bets.

34. Samuel Pepys wrote to Isaac Newton: "What is more likely, (a) at least one 6 in 6 rolls of one die or (b) at least two 6's in 12 rolls?" Compute the probabilities of these events.

Distributions

35. Suppose we roll two dice and let X and Y be the two numbers that appear. Find the distribution of $|X - Y|$.

36. Suppose we roll three tetrahedral dice that have 1, 2, 3, and 4 on their four sides. Find the distribution for the sum of the three numbers.

37. We roll two six-sided dice, one with sides 1, 2, 2, 3, 3, 4 and the other with sides 1, 3, 4, 5, 6, 8. What is the distribution of the sum?

38. How many children should a family plan to have so that the probability of having at least one child of each sex is ≥ 0.95?

39. How many times should a coin be tossed so that the probability of at least one head is $\geq 99\%$?

Expected value

40. You want to invent a gambling game in which a person rolls two dice and is paid some money if the sum is 7, but otherwise he loses his money. How much should you pay him for winning a $1 bet if you want this to be a fair game, that is, to have expected value 0?

41. A bet is said to carry 3 to 1 odds if you win $3 for each $1 you bet. What must the probability of winning be for this to be a fair bet?

42. A lottery has one $100 prize, two $25 prizes, and five $10 prizes. What should you be willing to pay for a ticket if 100 tickets are sold?

43. In a popular gambling game, three dice are rolled. For a $1 bet you win $1 for each 6 that appears (plus your dollar back). If no 6 appears you lose your dollar. What is your expected value?

44. A roulette wheel has slots numbered 1 to 36 and two labeled with 0 and 00. Suppose that all 38 outcomes have equal probabilities. Compute the expected values of the following bets. In each case you bet one dollar and when you win you get your dollar back in addition to your winnings. (a) You win $1 if one of the numbers 1 through 18 comes up. (b) You win $2 if the number that comes up is divisible by 3 (0 and 00 do not count). (c) You win $35 if the number 7 comes up.

45. In the Las Vegas game Wheel of Fortune, there are 54 possible outcomes. One is labeled "Joker," one "Flag," two "20," four "10," seven "5," fifteen "2," and twenty-four "1." If you bet $1 on a number you win that amount of money if the number comes up (plus your dollar back). If you bet $1 on Flag or Joker, you win $40 if that symbol comes up (plus your dollar back). What bets have the best and worst expected value here?

46. Sic Bo is an ancient Chinese dice game played with three dice. One of the possibilities for betting in the game is to bet "big." For this bet, you win if the total X is 11, 12, 13, 14, 15, 16, or 17, except when there are three 4's or three 5's. On a $1 bet on big, you win $1 plus your dollar back if it happens. What is your expected value?

47. Five people play a game of "odd man out" to determine who will pay for the pizza they ordered. Each flips a coin. If only one person gets heads (or tails) while the other four get tails (or heads) then he is the odd man and has to pay. Otherwise they flip again. What is the expected number of tosses needed to determine who will pay?

48. A man and his wife decide that they will keep having children until they have one of each sex. Ignoring the possibility of twins and supposing that each trial is independent and results in a boy or a girl with probability 1/2, what is the expected value of the number of children they will have?

49. An unreliable clothes dryer dries your clothes and takes 20 minutes with probability 0.6 and buzzes for 4 minutes and does nothing with probability 0.4. If we assume that successive trials are independent and that we patiently keep putting our money in to try to get it to work, what is the expected time we need to get our clothes dry?

Moments and variance

50. A random variable has $P(X = x) = x/15$ for $x = 1, 2, 3, 4, 5$, and 0 otherwise. Find the mean and variance of X.

51. Find the mean and variance of the number of games in the World Series. Recall that it is won by the first team to win four games and assume that the outcomes are determined by flipping a coin.

52. Suppose we pick a month at random from a non-leap year calendar and let X be the number of days in the month. Find the mean and variance of X.

53. The Elm Tree golf course in Cortland, NY, is a par 70 layout with 3 par fives, 5 par threes, and 10 par fours. Find the mean and variance of par on this course.

54. In a group of five items, two are defective. Find the distribution of N, the number of draws we need to find the first defective item. Find the mean and variance of N.

55. Can we have a random variable with $EX = 3$ and $EX^2 = 8$?

56. Suppose $P(X \in \{1, 2, 3\}) = 1$ and $EX = 2.5$. What are the smallest and largest possible values for the variance?

2

Combinatorial Probability

2.1 Permutations and combinations

As usual we begin with a question:

Example 2.1 The New York State Lottery picks 6 numbers out of 59, or more precisely, a machine picks 6 numbered ping-pong balls out of a set of 59. How many outcomes are there? The set of numbers chosen is all that is important. The order in which they are chosen is irrelevant.

To work up to the solution we begin with something that is obvious but is a key step in some of the reasoning to follow.

Example 2.2 A man has 4 pair of pants, 6 shirts, 8 pairs of socks, and 3 pairs of shoes. Ignoring the fact that some of the combinations may look ridiculous, in how many ways can he get dressed?

We begin by noting that there are $4 \cdot 6 = 24$ possible combinations of pants and shirts. Each of these can be paired with 1 of 8 choices of socks, so there are $24 \cdot 8 = 192$ ways of putting on pants, shirts, and socks. Repeating the last argument one more time, we see that for each of these 192 combinations there are 3 choices of shoes, so the answer is

$$4 \cdot 6 \cdot 8 \cdot 3 = 576 \text{ ways}$$

The reasoning in the last solution can clearly be extended to more than 4 experiments, and does not depend on the number of choices at each stage, so we have

The multiplication rule. Suppose that m experiments are performed in order and that, no matter what the outcomes of experiments $1, \ldots, k-1$ are, experiment k has n_k possible outcomes. Then the total number of outcomes is $n_1 \cdot n_2 \cdot \cdots \cdot n_m$.

Example 2.3

How many ways can 5 people stand in line?

To answer this question, we think about building the line up 1 person at a time starting from the front. There are 5 people we can choose to put at the front of the line. Having made the first choice, we have 4 possible choices for the second position. (The set of people we have to choose from depends on who was chosen first, but there are always 4 people to choose from.) Continuing, there are 3 choices for the third position, 2 for the fourth, and finally 1 for the last. Invoking the multiplication rule, we see that the answer must be

$$5 \cdot 4 \cdot 3 \cdot 2 \cdot 1 = 120$$

Generalizing from the last example we define n **factorial** to be

$$n! = n \cdot (n-1) \cdot (n-2) \cdots 2 \cdot 1 \tag{2.1}$$

To see that this gives the number of ways n people can stand in line, notice that there are n choices for the first person and $n-1$ for the second, and each subsequent choice reduces the number of people by 1 until finally there is only 1 person who can be the last in line.

Note that $n!$ grows very quickly since $n! = n \cdot (n-1)!$.

1!	1	7!	5,040
2!	2	8!	40,320
3!	6	9!	362,880
4!	24	10!	3,628,800
5!	120	11!	39,916,800
6!	720	12!	479,001,600

The number of ways we can put the 22 volumes of an encyclopedia on a shelf is

$$22! = 1.24000728 \times 10^{21}$$

Here, we have used our TI-83. We typed in 22 and then used the MATH button to get to the PRB menu and scroll down to the fourth entry to get the !, which gives us 22! after we press ENTER.

The number of ways that cards in a deck of 52 can be arranged is

$$52! = 8.065817517 \times 10^{67}$$

Before there were calculators, people used **Stirling's formula**:

$$n! \approx \left(\frac{n}{e}\right)^n \sqrt{2\pi n} \tag{2.2}$$

When $n = 52$, $52/e = 19.12973094$ and $\sqrt{2\pi n} = 18.07554591$, so

$$52! \approx (19.12973094)^{52} \cdot 18.07554591 = 8.0529 \times 10^{67}$$

Example 2.4

Twelve people belong to a club. How many ways can they pick a president, vice president, secretary, and treasurer?

Again we think of filling the offices one at a time in the order in which they were given in the last sentence. There are 12 people we can pick for president. Having made the first choice, there are always 11 possibilities for vice president, 10 for secretary, and 9 for treasurer. So by the multiplication rule, the answer is

$$\frac{12}{P}\ \frac{11}{V}\ \frac{10}{S}\ \frac{9}{T} = 11{,}800$$

To compute $P_{12,4}$ with the TI-83 calculator, type 12, push the MATH button, move the cursor across to the PRB submenu, scroll down to nPr on the second row, and press ENTER. nPr appears on the screen after the 12. Now type 4 and press ENTER.

Passing to the general situation, if we have k offices and n club members then the answer is

$$n \cdot (n-1) \cdot (n-2) \cdot \cdots \cdot (n-k+1)$$

To see this, note that there are n choices for the first office, $n-1$ for the second, and so on until there are $n-k+1$ choices for the last, since after the last person is chosen there will be $n-k$ left. Products such as the last one come up so often that they have a name: the **number of permutations of k objects from a set of size n**, or $P_{n,k}$ for short. Multiplying and dividing by $(n-k)!$, we have

$$n \cdot (n-1) \cdot (n-2) \cdot \cdots \cdot (n-k+1) \cdot \frac{(n-k)!}{(n-k)!} = \frac{n!}{(n-k)!}$$

which gives us a short formula

$$P_{n,k} = \frac{n!}{(n-k)!} \tag{2.3}$$

The last formula would give us the answer to the lottery problem if the order in which the numbers were drawn was important. Our last step is to consider a related but slightly simpler problem.

Example 2.5

A club has 23 members. How many ways can they pick 4 people to be on a committee to plan a party?

To reduce this question to the previous situation, we imagine making the committee members stand in line, which by (2.3) can be done in $23 \cdot 22 \cdot 21 \cdot 20$ ways. To get from this to the number of committees, we note that each committee can stand in line 4! ways, so the number of committees is the number of lineups divided by 4! or

$$\frac{23 \cdot 22 \cdot 21 \cdot 20}{1 \cdot 2 \cdot 3 \cdot 4} = 23 \cdot 11 \cdot 7 \cdot 5 = 8{,}855$$

To compute $C_{23,4}$ with the TI-83 calculator, type 23, push the MATH button, move the cursor across to the PRB submenu, scroll down to nCr on the third row, and press ENTER. nCr appears on the screen after the 23. Now type 4 and press ENTER.

Passing to the general situation, suppose we want to pick k people out of a group of n. Our first step is to make the k people stand in line, which can be done in $P_{n,k}$ ways, and then to realize that each set of k people can stand in line $k!$ ways, so the number of ways to choose k people out of n is

$$C_{n,k} = \frac{P_{n,k}}{k!} = \frac{n!}{k!(n-k)!} = \frac{n \cdot (n-1) \cdot \cdots \cdot (n-k+1)}{1 \cdot 2 \cdot \cdots \cdot k} \tag{2.4}$$

by (2.4) and (2.1). Here, $C_{n,k}$ is short for the **number of combinations of k things taken from a set of n**. $C_{n,k}$ is often written as $\binom{n}{k}$, a symbol that is read as "n choose k." We are now ready for the

Answer to the lottery problem, Example 2.1. We are choosing $k = 6$ objects out of a total of $n = 59$ when order is not important, so the number of possibilities is

$$C_{59,6} = \frac{59!}{6!53!} = \frac{59 \cdot 58 \cdot 57 \cdot 56 \cdot 55 \cdot 54}{1 \cdot 2 \cdot 3 \cdot 4 \cdot 5 \cdot 6}$$
$$= 59 \cdot 58 \cdot 19 \cdot 7 \cdot 11 \cdot 9 = 45{,}057{,}474$$

You should consider this the next time you think about spending \$1 for two chances to win a jackpot that starts at \$3 million and increases by \$1 million each week there is no winner.

Example 2.6 **World Series (continued).** Using (2.4) we can easily compute the probability that the series lasts 7 games. For this to occur the score must be tied 3–3 after 6 games. The total number of outcomes for the first 6 games is $2^6 = 64$. The number that ends in a 3–3 tie is

$$C_{6,3} = \frac{6!}{3!\,3!} = \frac{6 \cdot 5 \cdot 4}{1 \cdot 2 \cdot 3} = 20$$

since the outcome is determined by choosing the 3 games, team A will win. This gives us a probability of $20/64 = 5/16$ for the series to end in 7 games. Returning to the calculation in the previous section, we see that the number of outcomes that lead to A winning in 6 games is the number of ways of picking two of the first 5 games for B to win or $C_{5,2} = 5!/(2!\,3!) = 5 \cdot 4/2 = 10$.

Example 2.7 Suppose we flip 5 coins. Compute the probability that we get 0, 1, or 2 heads.

There are $2^5 = 32$ total outcomes. There is only 1 TTTTT that gives 0 head. If we want this to fit into our previous formula, we set $0! = 1$ (there is only one way for zero people to stand in line) so that

$$C_{5,0} = \frac{5!}{5!\,0!} = 1$$

There are 5 outcomes that have 1 head. We can see this by writing out the possibilities: HTTTT, THTTT, TTHTT, TTTHT, and TTTTH. Or, note that the number of ways to pick 1 toss for the heads to occur is

$$C_{5,1} = \frac{5!}{4!\,1!} = 5$$

Extending the last reasoning to 2 heads, the number of outcomes is the number of ways of picking 2 tosses for the heads to occur or

$$C_{5,2} = \frac{5!}{3!\,2!} = \frac{5 \cdot 4}{2} = 10$$

By symmetry the numbers of outcomes for 3, 4, and 5 heads are 10, 5, and 1, or in general.

$$C_{n,m} = C_{n,n-m} \tag{2.5}$$

The last equality is easy to prove: The number of ways of picking m objects out of n to take is the same as the number of ways of choosing $n - m$ to leave behind. Of course, one can also check this directly from the formula in (2.4).

Pascal's triangle. The number of outcomes for coin tossing problems fit together in a nice pattern:

```
                        1
                    1       1
                1       2       1
            1       3       3       1
        1       4       6       4       1
    1       5      10      10       5       1
 1      6      15      20      15       6       1
1   7      21      35      35      21       7       1
```

Each number is the sum of the 1's on the row above on its immediate left and right. To get the 1's on the edges to work correctly we consider the blanks to be 0's. In symbols,

$$C_{n,k} = C_{n-1,k-1} + C_{n-1,k} \tag{2.6}$$

Verbal proof. In picking k things out of n, which can be done in $C_{n,k}$ ways, we may or may not pick the last object. If we pick the last object then we must complete our set of k things by picking $k-1$ objects from the first $n-1$, which can be done in $C_{n-1,k-1}$ ways. If we do not pick the last object then we must pick all k objects from the first $n-1$, which can be done in $C_{n-1,k}$ ways. □

Analytic proof. Using the definition (2.4),

$$C_{n-1,k-1} + C_{n-1,k} = \frac{(n-1)!}{(n-k)!(k-1)!} + \frac{(n-1)!}{(n-k)!(k-1)!}$$

Factoring out the parts common to the two fractions

$$= \frac{(n-1)!}{(n-k-1)!(k-1)!}\left(\frac{1}{n-k} + \frac{1}{k}\right)$$

$$= \frac{(n-1)!}{(n-k-1)!(k-1)!}\left(\frac{n}{(n-k)k}\right) = \frac{n!}{(n-k)!k!}$$

which proves (2.6). □

Binomial theorem. The numbers in Pascal's triangle also arise if we take powers of $(x+y)$:

$$(x+y)^2 = x^2 + 2xy + y^2$$

$$(x+y)^3 = (x+y)(x^2+2xy+y^2) = x^3 + 3x^2y + 3xy^2 + y^3$$

$$(x+y)^4 = (x+y)(x^3+3x^2y+3xy^2+y^3)$$

$$= x^4 + 4x^3y + 6x^2y^2 + 4xy^3 + y^4$$

or in general

$$(x+y)^n = \sum_{m=0}^{n} C_{n,m} x^m y^{n-m} \tag{2.7}$$

To see this consider $(x+y)^5$ and write it as

$$(x+y)(x+y)(x+y)(x+y)(x+y)$$

Since we can choose x or y from each parenthesis, there are 2^5 terms in all. If we want a term of the form x^3y^2 then in 3 of the 5 cases we must pick x, so there are $C_{5,3} = (5\cdot 4)/2 = 10$ ways to do this. The same reasoning applies to the other terms, so we have

$$(x+y)^5 = C_{5,5}x^5 + C_{5,4}x^4y + C_{5,3}x^3y^2 + C_{5,2}x^2y^3 + C_{5,1}xy^4 + C_{5,0}y^5$$

$$= x^5 + 5x^4y + 10x^3y^2 + 10x^2y^3 + 5xy^4 + y^5$$

Poker. In the game of poker the following hands are possible; they are listed in increasing order of desirability. In the definitions the word *value* refers to A, K, Q, J, 10, 9, 8, 7, 6, 5, 4, 3, or 2. This sequence also describes the relative ranks of the cards, with one exception: an ace may be regarded as a 1 for the purposes of making a straight. (See the example in (d).)

(a) *One pair:* Two cards of equal value plus three cards with different values
J♠ J♢ 9♡ Q♣ 3♠
(b) *Two pair:* Two pairs plus another card with a different value
J♠ J♢ 9♡ 9♣ 3♠
(c) *Three of a kind:* Three cards of the same value and two with different values
J♠ J♢ J♡ 9♣ 3♠
(d) *Straight:* Five cards with consecutive values
5♡ 4♠ 3♠ 2♡ A♣
(e) *Flush:* Five cards of the same suit
K♣ 9♣ 7♣ 6♣ 3♣
(f) *Full house:* A three of a kind and a pair
J♠ J♢ J♡ 9♣ 9♠
(g) *Four of a kind:* Four cards of the same value plus another card
J♠ J♢ J♡ J♣ 9♠
(h) *Straight flush:* Five cards of the same suit with consecutive values
A♣ K♣ Q♣ J♣ 10♣

The last example is called a *royal flush.*

To compute the probabilities of these poker hands we begin by observing that there are

$$C_{52,5} = \frac{52 \cdot 51 \cdot 50 \cdot 49 \cdot 48}{1 \cdot 2 \cdot 3 \cdot 4 \cdot 5} = 2{,}598{,}960$$

ways of picking 5 cards out of a deck of 52, so it suffices to compute the number of ways each hand can occur. We will do three cases to illustrate the main ideas and then leave the rest to the reader:

(d) *Straight:* $10 \cdot 4^5$
A straight must start with a card that is 5 or higher, 10 possibilities. Once the values are decided on, suits can be assigned in 4^5 ways. This counting regards a straight flush as a straight. If you want to exclude straight flushes, suits can be assigned in $4^5 - 4$ ways.

(f) *Full house:* $13 \cdot C_{4,3} \cdot 12 \cdot C_{4,2}$
We first pick the value for the three of a kind (which can be done in 13 ways), then assign suits to those three cards ($C_{4,3}$ ways), then pick the value for the pair (12 ways), and then assign suits to the last two cards ($C_{4,2}$ ways).

(a) *One pair:* $13 \cdot C_{4,2} \cdot C_{12,3} \cdot 4^3$
We first pick the value for the pair (13 ways), next pick the suits for the pair ($C_{4,2}$ ways), and then pick three values for the other cards ($C_{12,3}$ ways) and assign suits to those cards (in 4^3 ways).

A common incorrect answer to this question is $13 \cdot C_{4,2} \cdot 48 \cdot 44 \cdot 40$. The faulty reasoning underlying this answer is that the third card must not have the same value as the cards in the pair (48 choices), the fourth must be different from the third and the pair (44 choices), ... However, this reasoning is flawed since it counts each outcome $3! = 6$ times. (Note that $48 \cdot 44 \cdot 40/3! = C_{12,3} \cdot 4^3$.)

The numerical values of the probabilities of all poker hands are given in the next table.

(a) *One pair*	0.422569
(b) *Two pair*	0.047539
(c) *Three of a kind*	0.021128
(d) *Straight*	0.003940
(e) *Flush*	0.001981
(f) *Full house*	0.001441
(g) *Four of a kind*	0.000240
(h) *Straight flush*	0.000015

The probability of getting none of these hands can be computed by summing the values for (a) through (g) (recall that (d) includes (h)) and subtracting the result from 1. However, it is much simpler to observe that we have nothing if we have 5 different values that do not make a straight or a flush. So the number of nothing hands is $(C_{13,5} - 10) \cdot (4^5 - 4)$ and the probability of a nothing hand is 0.501177.

2.1.1 More than two categories

We defined $C_{n,k}$ as the number of ways of picking k objects out of n. To motivate the next generalization we would like to observe that $C_{n,k}$ is also the number of ways we can divide n objects into two groups, the first one with k objects and the second with $n - k$. To connect this observation with the next problem, think of it as asking: "How many ways can we divide 12 objects into three numbered groups of sizes 4, 3, and 5?"

Example 2.8 A house has 12 rooms. We want to paint 4 yellow, 3 purple, and 5 red. In how many ways can this be done?

This problem can be solved using what we know already. We first pick 4 of the 12 rooms to be painted yellow, which can be done in $C_{12,4}$ ways, and then pick 3 of the remaining 8 rooms to be painted purple, which can be done in $C_{8,3}$ ways. (The 5 unchosen rooms will be painted red.) The answer is

$$C_{12,4}C_{8,3} = \frac{12!}{4!\,8!} \cdot \frac{8!}{3!\,5!} = \frac{12!}{4!\,3!\,5!} = 27{,}720$$

A second way of looking at the problem, which gives the last answer directly, is to first decide the order in which the rooms will be painted, which can be done in 12! ways, and then paint the first 4 on the list yellow, the next 3 purple, and the last 5 red. One example is

$$\begin{array}{cccccccccccc} 9 & 6 & 11 & 1 & 8 & 2 & 10 & 5 & 3 & 7 & 12 & 4 \\ \hline \text{Y} & \text{Y} & \text{Y} & \text{Y} & \text{P} & \text{P} & \text{P} & \text{R} & \text{R} & \text{R} & \text{R} & \text{R} \end{array}$$

Now, the first four choices can be rearranged in 4! ways without affecting the outcome, the middle three in 3! ways, and the last five in 5! ways. Invoking the multiplication rule, we see that in a list of the 12! possible permutations each possible painting thus appears 4! 3! 5! times. Hence the number of possible paintings is

$$\frac{12!}{4!\,3!\,5!}$$

The second computation is a little more complicated than the first, but makes it easier to see

Theorem 2.1. *The number of ways a group of n objects to be divided into m groups of size $n_1, \ldots, n_m$ with $n_1 + \cdots + n_m = n$ is*

$$\frac{n!}{n_1!\,n_2! \cdots n_m!} \tag{2.8}$$

The formula may look complicated but it is easy to use.

Example 2.9

There are 39 students in a class. In how many ways can a professor give out 9 A's, 13 B's, 12 C's, and 5 F's?

$$\frac{39!}{9!\,13!\,12!\,5!} = 1.57 \times 10^{22}$$

Example 2.10

Bridge. Four people play a card game in which each gets 13 cards. How many possible deals are there?

$$\frac{52!}{(13!)^4} = 5.364473777 \times 10^{28}$$

Example 2.11

Suppose we draw 13 cards from a deck. How many outcomes are there? How many lead to hands with 4 spades, 3 hearts, 3 diamonds, and 3 clubs? 3 spades, 5 hearts, 2 diamonds, and 3 clubs?

$$C_{52,13} = 6.350135596 \times 10^{11}$$

$$C_{13,4}C_{13,3}C_{13,3}C_{13,3} = 715 \cdot (286)^3 = 16{,}726{,}464{,}040$$

$$C_{13,3}C_{13,5}C_{13,2}C_{13,3} = 286 \cdot 1{,}287 \cdot 78 \cdot 286 = 8{,}211{,}173{,}256$$

Example 2.12

Suit distributions. The last bridge hand in the previous example is said to have a 5–3–3–2 distribution. Here, we have listed the number cards in the longest suit first and continued in decreasing order. Permuting the four numbers we see that the example 3 spades, 5 hearts, 2 diamonds, and 3 clubs is just one of $4!/2!$ possible ways of assigning the numbers to suits, so the probability of a 5–3–3–2 distribution is

$$\frac{12 \cdot 8{,}211{,}173{,}256}{6.350135596 \times 10^{11}} = 0.155$$

Similar computations lead to the results in the next table. We have included the number of different permutations of the pattern to help explain why slightly unbalanced distributions have larger probability than 4–3–3–3.

Distribution	Probability	Permutations
4–4–3–2	0.216	12
5–3–3–2	0.155	12
5–4–3–1	0.129	24
5–4–2–2	0.106	12
4–3–3–3	0.105	4
6–3–2–2	0.056	12

2.2 Binomial and multinomial distributions

Example 2.13

Suppose we roll 6 dice. What is the probability of $A =$ "we get exactly two 4's"?

One way that A can occur is

$$\frac{\times}{1}\ \frac{4}{2}\ \frac{\times}{3}\ \frac{4}{4}\ \frac{\times}{5}\ \frac{\times}{6}$$

where $\times$ stands for "not a 4." Since the six events "die 1 shows $\times$," "die 2 shows 4," ..., "die 6 shows $\times$" are independent, the indicated pattern has probability

$$\frac{5}{6} \cdot \frac{1}{6} \cdot \frac{5}{6} \cdot \frac{1}{6} \cdot \frac{5}{6} \cdot \frac{5}{6} = \left(\frac{1}{6}\right)^2 \left(\frac{5}{6}\right)^4$$

Here, we have been careful to say "pattern" rather than "outcome" since the given pattern corresponds to 5^4 outcomes in the sample space of 6^6 possible outcomes for 6 dice. Each pattern that results in A corresponds to a choice of 2 of the 6 trials on which a 4 will occur, so the number of patterns is $C_{6,2}$. When we write out the probability of each pattern, there will be two 1/6s and four 5/6s, so each pattern has the same probability and

$$P(A) = C_{6,2}\left(\frac{1}{6}\right)^2\left(\frac{5}{6}\right)^4$$

Generalizing from the last example, suppose we perform an experiment n times and on each trial an event we call "success" has probability p. (Here and in what follows, when we repeat an experiment, we assume that the outcomes of the various trials are independent.) Then the probability of k successes is

$$C_{n,k}\,p^k(1-p)^{n-k} \tag{2.9}$$

This is called the **binomial(n, p) distribution**. Taking $n = 6$, $k = 2$, and $p = 1/6$ in (2.9) gives the answer in the previous example. The reasoning for the general formula is similar. There are $C_{n,k}$ ways of picking k of the n trials for successes to occur, and each pattern of k successes and $n - k$ failures has probability $p^k(1-p)^{n-k}$.

Theorem 2.2. *The binomial(n, p) distribution has mean np and variance $np(1-p)$.*

Proof using theory. Let $X_i = 1$ if the ith trial is a success and 0 otherwise. $S_n = X_1 + \cdots + X_n$ is the number of successes in n trials. Using (1.8) we see that $E S_n = nE X_i = np$; that is, the expected number of successes is the number of trials n times the success probability p on each trial.

Since $X_1, \ldots, X_n$ are independent, (6.11) implies that $\text{var}(S_n) = \text{var}(X_1) + \cdots + \text{var}(X_n) = n\,\text{var}(X_1)$. To compute $\text{var}(X_1)$, we note that $EX_1^2 = 1 \cdot p + 0 \cdot (1-p) = p$, so $\text{var}(X_1) = EX_1^2 - (EX_1)^2 = p - p^2 = p(1-p)$ and the desired result follows. □

Proof by computation. Using the definition of expected value,

$$EN = \sum_{m=0}^{n} m\frac{n!}{m!(n-m)!}p^m(1-p)^{n-m}$$

The $m = 0$ term contributes nothing, so we can cancel m's and rearrange to get

$$np \sum_{m=1}^{n} m \frac{(n-1)!}{(m-1)!(n-m)!} p^{m-1}(1-p)^{n-m} = np$$

since the sum computes the total probability for the binomial$(n-1, p)$ distribution.

As in the case of the geometric, our next step is to compute

$$E(N(N-1) = \sum_{m=2}^{n} m(m-1) \frac{n!}{m!(n-m)!} p^m (1-p)^{n-m}$$

We have dropped the first two terms that contribute nothing, so we can cancel to get

$$n(n-1)p^2 \sum_{m=2}^{n} m(m-1) \frac{(n-2)!}{(m-2)!(n-m)!} p^{m-2}(1-p)^{n-m}$$

since the sum computes the total probability for the binomial$(n-2, p)$ distribution.

To finish up, we note that

$$\begin{aligned} \text{var}(N) &= EN^2 - (EN)^2 = E(N(N-1)) = EN - (EN)^2 \\ &= n(n-1)p^2 + np - n^2p^2 = n(p - p^2) = np(1-p) \end{aligned}$$

which completes the proof. □

Example 2.14 A student takes a test with 16 multiple-choice questions. Since she has never been to class she has to choose at random from the 4 possible answers. What is the probability that she will get exactly 3 right?

The number of trials is $n = 10$. Since she is guessing the probability of success $p = 1/4$, so using (2.9) the probability of $k = 3$ successes and $n - k = 7$ failures is

$$C_{16,3}(1/4)^3(3/4)^7 = \frac{16 \cdot 15 \cdot 14}{1 \cdot 2 \cdot 3} \cdot \frac{3^{13}}{4^{16}} = 560 \cdot \frac{1{,}594{,}323}{4{,}294{,}967{,}296} = 0.2079$$

In the same way we can compute the other probabilities. The results are given in next figure. With a TI-83 calculator these answers can be found by going to the DISTR menu and using binompdf$(16, 0.25, k)$. Here, pdf is short for probability density function.

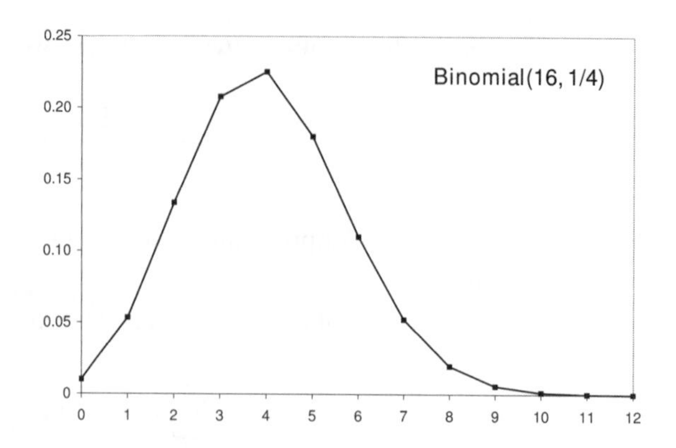

Example 2.15 A football team wins each week with probability 0.6 and loses with probability 0.4. If we suppose that the outcomes of their 10 games are independent, what is the probability that they will win exactly 8 games?

The number of trials is $n = 10$. We are told that the success probability $p = 0.6$, so by (2.9), the probability of $k = 8$ successes and $n - k = 2$ failures is

$$C_{10,8}(0.6)^8(0.4)^2 = \frac{10 \cdot 9}{1 \cdot 2}(0.6)^8(0.4)^2 = 0.1209$$

In the same way we can compute the other probabilities:

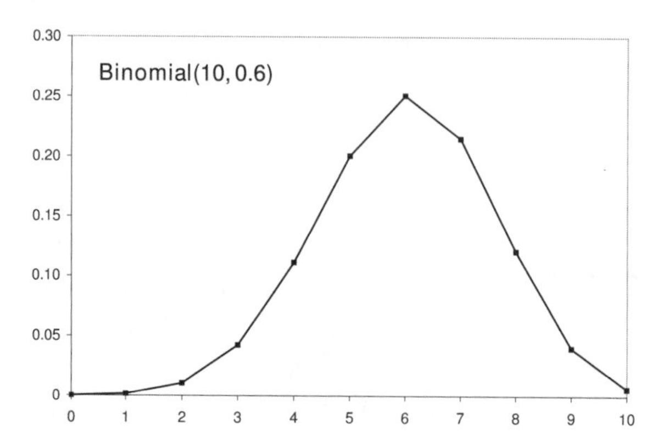

Example 2.16 **Aces at bridge.** When we draw 13 cards out of a deck of 52, each ace has a probability 1/4 of being chosen, but the four events are not independent. How does the probability of $k = 0, 1, 2, 3, 4$ aces compare with that of the binomial distribution with $n = 4$ and $p = 1/4$?

We first consider the probability of drawing two aces:

$$\frac{C_{4,2}C_{48,11}}{C_{52,13}} = \frac{6 \cdot \frac{48 \cdots 38}{11!}}{\frac{52 \cdots 40}{13!}} = 6 \cdot \frac{13 \cdot 12 \cdot 39 \cdot 38}{52 \cdot 51 \cdot 50 \cdot 49} = 0.2135$$

In contrast, the probability for the binomial is

$$C_{4,2}(1/4)^2(3/4)^2 = 0.2109$$

To compare the two formulas, note that $13/52 = 1/4$, $12/51 = 0.2352$, $39/50 = 0.78, 38/51 = 0.745$ versus $(1/4)^2(3/4)^2$ in the binomial formula. Similar computations show that if $D = 52 \cdot 51 \cdot 50 \cdot 49$, then the answers are

	Aces	Binomial
0	$\frac{39 \cdot 38 \cdot 37 \cdot 36}{52 \cdot 51 \cdot 50 \cdot 49}$	$(3/4)^4$
1	$4 \cdot \frac{13 \cdot 39 \cdot 38 \cdot 37}{52 \cdot 51 \cdot 50 \cdot 49}$	$4 \cdot (1/4)(3/4)^3$
2	$6 \cdot \frac{13 \cdot 12 \cdot 39 \cdot 38}{52 \cdot 51 \cdot 50 \cdot 49}$	$6 \cdot (1/4)^2(3/4)^2$
3	$4 \cdot \frac{13 \cdot 12 \cdot 11 \cdot 39}{52 \cdot 51 \cdot 50 \cdot 49}$	$4 \cdot (1/4)^3(3/4)$
4	$\frac{13 \cdot 12 \cdot 11 \cdot 10}{52 \cdot 51 \cdot 50 \cdot 49}$	$(1/4)^4$

Evaluating these expressions leads to the following probabilities:

	Aces	Binomial
0	0.3038	0.3164
1	0.4388	0.4218
2	0.2134	0.2109
3	0.0412	0.0468
4	0.00264	0.00390

Example 2.17

In 8 games of bridge, Harry had 6 hands without an ace. Should he doubt that the cards are being shuffled properly?

The number of hands with no ace has a binomial distribution with $n = 8$ and $p = 0.3038$. The probability of at least 6 hands without an ace is

$$\sum_{k=6}^{8} C_{8,k}(0.3038)^k(0.6962)^{8-k} = 1 - \sum_{k=0}^{5} C_{8,k}(0.3038)^k(0.6962)^{8-k}$$

We have turned the probability around because on the TI-83 calculator the sum can be evaluated as binomcdf(8, 0.3038, 5) = 0.9879. Here, cdf is short for cumulative distribution function, that is, the probability of ≤ 5 hands without an ace. Thus the probability of luck this bad is 0.0121.

Example 2.18

Tennis. In men's tennis, the winner is the first to win 3 out of 5 sets. If Roger Federer wins a set against his opponent with probability 2/3, what is the probability w that he will win the match?

He can win in three sets, four or five, but he must win the last set, so

$$w = \left(\frac{2}{3}\right)^3 + C_{3,2}\left(\frac{2}{3}\right)^3\frac{1}{3} + C_{4,2}\left(\frac{2}{3}\right)^3\left(\frac{1}{3}\right)^2$$

$$= \left(\frac{2}{3}\right)^3 (1 + 3(1/3) + 6(1/9)) = \frac{8}{27}\cdot\frac{8}{3} = 0.790$$

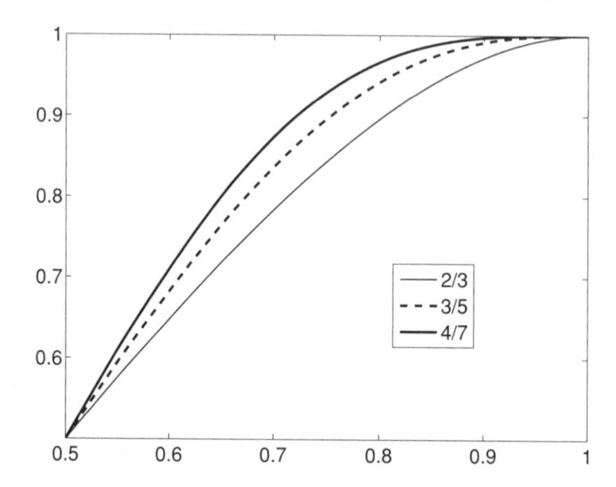

Replacing 2/3 by p and 1/3 by $(1-p)$, we get the general solution for best 3 out of 5, which generalizes easily to other common formats:

2 out of 3: $p^2 + C_{2,1}p^2(1-p)$
3 out of 5: $p^3 + C_{3,2}p^3(1-p) + C_{4,2}p^3(1-p)^2$
4 out of 7: $p^4 + C_{4,3}p^4(1-p) + C_{5,3}p^4(1-p)^2 + C_{6,3}p^4(1-p)^3$

As we should expect, if $p > 1/2$, then the winning probability increases with the length of the series. When $p = 0.6$ the three values are 0.648, 0.68256, and 0.710208. The graph above compares the winning probabilities for a team that wins each game with probability p.

Multinomial distribution

The arguments above generalize easily to independent events with more than two possible outcomes. We begin with an example.

Example 2.19

Consider a die with 1 painted on three sides, 2 painted on two sides, and 3 painted on one side. If we roll this die 10 times, what is the probability that we get five 1's, three 2's and two 3's?

The answer is

$$\frac{10!}{5!\,3!\,2!}\left(\frac{1}{2}\right)^5\left(\frac{1}{3}\right)^3\left(\frac{1}{6}\right)^2$$

The first factor, by (2.8), gives the number of ways to pick five rolls for 1's, three rolls for 2's, and two rolls for 3's. The second factor gives the probability of any outcome with five 1's, three 2's, and two 3's. Generalizing from this example, we see that if we have k possible outcomes for our experiment with probabilities $p_1, \ldots, p_k$, then the probability of getting exactly n_i outcomes of type i in $n = n_1 + \cdots + n_k$ trials is

$$\frac{n!}{n_1! \cdots n_k!}\, p_1^{n_1} \cdots p_k^{n_k} \tag{2.10}$$

since the first factor gives the number of outcomes and the second the probability of each one.

Example 2.20 A baseball player gets a hit with probability 0.3, a walk with probability 0.1, and an out with probability 0.6. If he bats 4 times during a game and we assume that the outcomes are independent, what is the probability that he will get 1 hit, 1 walk, and 2 outs?

The total number of trials is $n = 4$. There are $k = 3$ categories: hit, walk, and out. $n_1 = 1$, $n_2 = 1$, and $n_3 = 2$. Plugging in to our formula the answer is

$$\frac{4!}{1!1!2!}(0.3)(0.1)(0.6)^2 = 0.1296$$

Example 2.21 The output of a machine is graded excellent 70% of the time, good 20% of the time, and defective 10% of the time. What is the probability that a sample of size 15 has 10 excellent, 3 good, and 2 defective items?

The total number of trials is $n = 15$. There are $k = 3$ categories: excellent, good, and defective. We are interested in outcomes with $n_1 = 10$, $n_2 = 3$, and $n_3 = 2$. Plugging in to our formula the answer is

$$\frac{15!}{10!\,3!\,2!} \cdot (0.7)^{10}(0.2)^3(0.1)^2$$

2.3 Poisson approximation to the binomial

X is said to have a **Poisson distribution** with parameter λ, or Poisson(λ) if

$$P(X = k) = e^{-\lambda}\frac{\lambda^k}{k!} \quad \text{for } k = 0, 1, 2, \ldots$$

Here, $\lambda > 0$ is a parameter. To see that this is a probability function we recall

$$e^x = \sum_{k=0}^{\infty} \frac{x^k}{k!} \tag{2.11}$$

so the proposed probabilities are nonnegative and sum to 1.

The next figure shows the Poisson distribution with $\lambda = 4$.

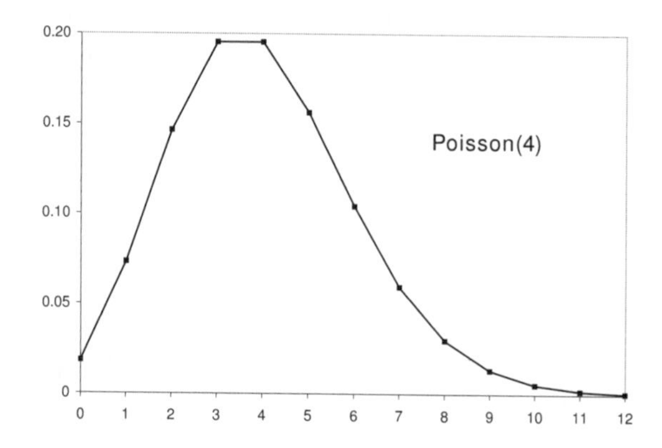

Theorem 2.3. *The Poisson distribution has mean λ and variance λ.*

Proof. Since the $k = 0$ term makes no contribution to the sum,

$$EX = \sum_{k=1}^{\infty} ke^{-\lambda}\frac{\lambda^k}{k!} = \lambda \sum_{k=1}^{\infty} e^{-\lambda}\frac{\lambda^{k-1}}{(k-1)!} = \lambda$$

since $\sum_{k=1}^{\infty} P(X = (k-1)) = 1$. As in the case of the binomial and geometric our next step is to compute

$$E(X(X-1)) = \sum_{k=2}^{\infty} k(k-1)e^{-\lambda}\frac{\lambda^k}{k!} = \lambda^2 \sum_{k=2}^{\infty} e^{-\lambda}\frac{\lambda^{k-2}}{(k-2)!} = \lambda^2$$

From this it follows that

$$\begin{aligned} \text{var}\,(X) &= EX^2 - (EX)^2 = E(X(X-1)) + EX - (EX)^2 \\ &= \lambda^2 + \lambda - \lambda^2 \end{aligned}$$

which completes the proof. □

Our next result explains why the Poisson distribution arises in a number of situations.

Theorem 2.4. *Suppose S_n has a binomial distribution with parameters n and p_n. If $p_n \to 0$ and $np_n \to \lambda$ as $n \to \infty$, then*

$$P(S_n = k) \to e^{-\lambda} \frac{\lambda^k}{k!} \tag{2.12}$$

In words, if we have a large number of independent events with small probability then the number that occurs has approximately a Poisson distribution. The key to the proof is the following fact.

Lemma. *If $p_n \to 0$ and $np_n \to \lambda$, then as $n \to \infty$*

$$(1 - p_n)^n \to e^{-\lambda} \tag{2.13}$$

Proof. Calculus tells us that if x is small then

$$\ln(1 - x) = -x - \frac{x^2}{2} - \cdots$$

Using this we have

$$\begin{aligned}(1 - p_n)^n &= \exp(n \ln(1 - p_n)) \\ &\approx \exp(-np_n - np_n^2/2) \approx \exp(-\lambda)\end{aligned}$$

In the last step we used the observation that $p_n \to 0$ to conclude that $np_n \cdot p_n/2$ is much smaller than np_n. □

Proof of (2.12). Since $P(S_n = 0) = (1 - p_n)^n$, (2.13) gives the result for $k = 0$. To prove the result for $k > 0$, we let $\lambda_n = np_n$ and observe that

$$\begin{aligned}P(S_n = k) &= C_{n,k} \left(\frac{\lambda_n}{n}\right)^k \left(1 - \frac{\lambda_n}{n}\right)^{n-k} \\ &= \frac{n(n-1)\cdots(n-k+1)}{n^k} \cdot \frac{\lambda_n^k}{k!} \left(1 - \frac{\lambda_n}{n}\right)^n \left(1 - \frac{\lambda_n}{n}\right)^{-k} \\ &\to 1 \cdot \frac{\lambda^k}{k!} \cdot e^{-\lambda} \cdot 1\end{aligned}$$

Here, $n(n-1)\cdots(n-k+1)/n^k \to 1$ since there are k factors in the numerator, and for each fixed j, $(n-j)/n = 1 - (j/n) \to 1$. The last term $(1 - \{\lambda_n/n\})^{-k} \to 1$ since k is fixed and $1 - \{\lambda_n/n\} \to 1$. □

When we apply (2.12), we think, "If $S_n = \text{binomial}(n, p)$ and p is small then S_n is approximately Poisson(np)." The next example illustrates the use of this approximation and shows that the number of trials does not have to be very large for us to get accurate answers.

Example 2.22

Suppose we roll two dice 12 times and we let D be the number of times a double 6 appears. Here, $n = 12$ and $p = 1/36$, so $np = 1/3$. We now compare $P(D = k)$ with the Poisson approximation for $k = 0, 1, 2$.

$$k = 0 \text{ exact answer:} \quad P(D = 0) = \left(1 - \frac{1}{36}\right)^{12} = 0.7132$$

$$\text{Poisson approximation:} \quad P(D = 0) = e^{-1/3} = 0.7165$$

$$k = 1 \text{ exact answer:} \quad P(D = 1) = C_{12,1}\frac{1}{36}\left(1 - \frac{1}{36}\right)^{11}$$

$$= \left(1 - \frac{1}{36}\right)^{11} \cdot \frac{1}{3} = 0.2445$$

$$\text{Poisson approximation:} \quad P(D = 1) = e^{-1/3}\frac{1}{3} = 0.2388$$

$$k = 2 \text{ exact answer:} \quad P(D = 2) = C_{12,2}\left(\frac{1}{36}\right)^{2}\left(1 - \frac{1}{36}\right)^{10}$$

$$= \left(1 - \frac{1}{36}\right)^{10} \cdot \frac{12 \cdot 11}{36^2} \cdot \frac{1}{2!} = 0.0384$$

$$\text{Poisson approximation:} \quad P(D = 2) = e^{-1/3}\left(\frac{1}{3}\right)^{2}\frac{1}{2!} = 0.0398$$

Example 2.23

Death by horse kick. Ladislaus Bortkiewicz published a book about the Poisson distribution titled *The Law of Small Numbers* in 1898. In this book he analyzed the number of German soldiers kicked to death by cavalry horses between 1875 and 1894 in each of 14 cavalry corps, arguing that it fits the Poisson distribution. I. J. Good and others have argued that the Poisson distribution should be called the Bortkiewicz distribution, but then it would be very difficult to say or write.

Example 2.24

V-2 rocket hits in London during World War II. The area under study was divided into 576 areas of equal size. There were a total of 537 hits or an average of 0.9323 per subdivision. Using the Poisson distribution the probability a subdivision is not hit is $e^{-0.9323} = 0.3936$. Multiplying by 576 we see that the expected number not hit was 226.71, which agrees well with the 229 that were observed not to be hit.

Example 2.25

Shark attacks. In the summer of 2001 there were 6 shark attacks in Florida, while the yearly average is 2. Is this unusual?

In an article in the September 7, 2001, *National Post*, Professor David Kelton of Penn State University argued that this was a random event. "Just because you see events happening in a rash this does not imply that there is some physical driver causing them to happen. It is characteristic of random processes that they have bursty behavior." He did not seem to realize that the probability of six shark attacks under the Poisson distribution is

$$e^{-2}\frac{2^6}{6!} = 0.01203$$

This probability can be found with the TI-83 by using Poissonpdf(2, 6) on DISTR menu. If we want the probability of at least 6, we would use 1 − Poissoncdf(2, 5).

Example 2.26

Alliteration in Shakespeare. Did Shakespeare consciously choose words with the same sounds or did lines like "full fathom five thy father lies" just occur by chance. Psychologist B. F. Skinner addressed this question in two papers (one in 1939 in *The Psychological Record*, Vol. 3, pp. 186–192, and one in 1941 in *The American Journal of Psychology*, Vol. 54, pp. 64–79). He looked at 100 sonnets (for a total of 1,400 lines) and counted the number of times s sounds appeared in a line. The next table compares the counts to those expected under a Poisson distribution with the same mean.

s sounds	0	1	2	3	4
Observed	702	501	161	29	7
Expected	685	523	162	26	2

It turns out that most of the discrepancy in the last two cells goes away if the same word on a line is not counted more than once. However, even without this manipulation the similarity to the Poisson is remarkable. This example comes from a 1989 article by Diaconis and Mosteller on coincidences in the *Journal of the American Statistical Association*, Vol. 84, pp. 853–861.

Example 2.27

Birthday problem, II. If we are in a group of $n = 183$ individuals, what is the probability that no one else has our birthday?

By (2.13), the probability is

$$\left(1 - \frac{1}{365}\right)^{182} \approx e^{-182/365} = 0.6073$$

From this we see that in order to have a probability of about 0.5, we need $365 \ln 2 = 253$ people as we calculated before.

Example 2.28

Birthday problem, I. Consider now a group of $n = 25$ and ask our original question: What is the probability that two people have the same birthday?

The events $A_{i,j}$ that persons i and j have the same birthday are only pairwise independent, so strictly speaking (2.12) does not apply. However, it gives a reasonable approximation. The number of pairs of people is $C_{25,2} = 300$, while the probability of a match for a given pair is 1/365, so by (2.13) the probability of no match is

$$\approx \exp(-300/365) = 0.4395$$

versus the exact probability of 0.4313 from the table in Section 1.1.

General Poisson approximation result. The Poisson distribution is often used as a model for the number of people who go to a fast-food restaurant between 12 and 1, the number of people who make a cell phone call between 1:45 and 1:50, or the number of traffic accidents in a day. To explain the reasoning in the last case we note that any one person has a small probability of having an accident on a given day, and it is reasonable to assume that the events A_i = "the ith person has an accident" are independent. However, it is not reasonable to assume that the probabilities of having an accident $p_i = P(A_i)$ are all the same, nor is it reasonable to assume that all women have the same probability of giving birth, but fortunately the Poisson approximation does not require this.

Theorem 2.5. *Consider independent events $A_i, i = 1, 2, \ldots, n$, with probabilities $p_i = P(A_i)$. Let N be the number of events that occur, let $\lambda = p_1 + \cdots + p_n$, and let Z have a Poisson distribution with parameter λ. Then, for any set of integers B,*

$$|P(N \in B) - P(Z \in B)| \leq \sum_{i=1}^{n} p_i^2 \tag{2.14}$$

We can simplify the right-hand side by noting

$$\sum_{i=1}^{n} p_i^2 \leq \max_i p_i \sum_{i=1}^{n} p_i = \lambda \max_i p_i$$

This says that if all the p_i are small then the distribution of N is close to a Poisson with parameter λ. Taking $B = \{k\}$, we see that the individual probabilities $P(N = k)$ are close to $P(Z = k)$, but this result says more. The probabilities of events such as $P(3 \leq N \leq 8)$ are close to $P(3 \leq Z \leq 8)$ and we have an explicit bound on the error.

For a concrete situation, consider Example 2.22, where $n = 12$ and all the $p_i = 1/36$. In this case the error bound is

$$\sum_{i=1}^{12} p_i^2 = 12\left(\frac{1}{36}\right)^2 = \frac{1}{108} = 0.00926$$

while the error for the approximation for $k = 1$ is 0.0057.

Example 2.29

The previous example justifies the use of the Poisson distribution in modeling the number of visits to a Web site in a minute. Suppose that the average number of visitors per minute is $\lambda = 5$, but that the site will crash if there are 12 visitors or more. What is the probability that the site will crash?

The probability of 12 or more visitors is

$$1 - \sum_{k=0}^{11} e^{-5} \frac{5^k}{k!}$$

This would be tedious to do by hand, but is easy if we use the TI-83 calculator. Using the distributions menu,

$$\sum_{k=0}^{11} e^{-5} \frac{5^k}{k!} = \text{Poissoncdf}(5, 11) = 0.994547$$

Subtracting this from 1, we have the answer 0.005453.

Example 2.30

Births in Ithaca. The Poisson distribution can be used for births as well as for deaths. There were 63 births in Ithaca, NY, between March 1 and April 8, 2005, a total of 39 days, or 1.615 per day. The next table gives the observed number of births per day and compares with the prediction from the Poisson distribution.

	0	1	2	3	4	5	6
Observed	9	12	9	5	3	0	1
Poisson	7.75	12.52	10.11	5.44	2.19	0.71	0.19

Example 2.31

Wayne Gretsky. He scored a remarkable 1,669 points in 696 games as an Edmonton Oiler, for a rate of $1{,}669/696 = 2.39$ points per game. From the Poisson formula with $k = 0$, the probability of Gretzky having a pointless game is $e^{-2.39} = 0.090$. The next table compares that actual number of games with the numbers predicted by the Poisson approximation.

Points	Games	Poisson
0	69	63.27
1	155	151.71
2	171	181.90
3	143	145.40
4	79	87.17
5	57	41.81
6	14	16.71
7	6	5.72
8	2	1.72
9	0	0.46

Coincidences. If we see an event with a chance of one in a hundred million then we are amazed, even though each day in the United States such events will happen to three people. This observation helps explain why some incredible things occur. Some times, as in our first example, it is just a miscalculation.

Example 2.32

Pick 4 coincidence. To quote a United Press story on September 10, 1981, reported by James Hanley in his article in the *American Statistician*, Vol. 46, pp. 197–202: "Lottery officials say that there is 1 chance in 100 million that the same four digit lottery number would be drawn in Massachusetts and New York on the same night. That's just what happened Tuesday. The number 8902 came up paying $5842 in Massachusetts and $4,500 New York."

These lotteries pick four-digit numbers, so each number has a 1 in 10^4 chance. The probability that 8902 is chosen in both states is 1 in 10^8 but then again some number will be chosen in Massachusetts and after that has been done the probability that the same number is chosen in New York is 1 in 10^4. When you take into account that there are a half dozen states that have similar games and drawings occur twice a day in New York, one should expect this to happen every few years.

Example 2.33

Lottery double winner. The following item was reported in the February 14, 1986, edition of the *New York Times*: A New Jersey woman, Evelyn Adams, won the lottery twice within a span of 4 months raking in a total of 5.4 million dollars. She won the jackpot for the first time on October 23, 1985, in the Lotto 6/39, in which you pick 6 numbers out of 39. Then she won the jackpot in the new Lotto 6/42 on February 13, 1986. Lottery officials calculated the probability of this as roughly one in 17.1 trillion. What do you think of this statement?

It is easy to see where they get this from. The probability of a person picked in advance of the lottery getting all six numbers right both times is

$$\frac{1}{C_{39,6}} \cdot \frac{1}{C_{42,6}} = \frac{1}{17.1 \times 10^{12}}$$

One can immediately reduce the odds against this event by noting that the first lottery had some winner, who if they played only one ticket in the second lottery had a $1/C_{42,6}$ chance.

The odds drop even further when you consider that there are a large number of people who submit more than one entry for the twice weekly drawing and that wins on October 23, 1985, and February 13, 1986, is not the only combination. Suppose for concreteness that each week 1 million people play the lottery and each buys exactly five tickets. The probability of one person winning on a given week is

$$p_1 = \frac{5}{C_{42,6}} = 9.531 \times 10^{-7}$$

The number of times one person will win a jackpot in the next year (100 twice-weekly drawings) is roughly Poisson with mean

$$\lambda_1 = 100 p_1 = 9.531 \times 10^{-5}$$

The probability that a given player wins the jackpot two or more times is

$$p_0 = 1 - e^{-\lambda_1} - e^{-\lambda_1}\lambda_1 = 4.54 \times 10^{-9}$$

The number of double winners in a population of 1 million players is Poisson with mean

$$\lambda_0 = (1{,}000{,}000)\, p_0 = 4.54 \times 10^{-3}$$

so the probability of at least one double winner is $1 - e^{-0.00454} \approx 0.00454$. If you take into account that many states have lotteries and we were just looking at 1 year, we see that a double winner is not unusual at all.

My favorite double-winner story is Maureen Wilcox. In June 1980, she bought tickets for both the Massachusetts Lottery and the Rhode Island Lottery. She picked the winning numbers for both lotteries. Unfortunately for her, her Massachusetts numbers won in Rhode Island and vice versa.

Example 2.34

Scratch-off triple winner. 81-year-old Keith Selix won three lottery prizes totaling \$81,000 from scratch-off games in the 12 months preceding May 3, 2006. He won \$30,00 twice in "Wild Crossword" games and \$21,000 playing "Double Blackjack." Again we want to calculate the probability of this.

The odds of winning in these games are 89,775 to 1 and 119,700 to 1 respectively. One of the reasons Selix won so many times in 2006 is that he spent about \$200 a week or more than \$10,000 a year on scratch-off games. If the games cost \$1 then this would be 10,000 plays with an approximate 1/100,000 chance of winning. Thus his expected number of wins would be 0.1 and the probability of exactly three wins would be

$$e^{-0.1}\frac{(0.1)^3}{3!} \quad \text{or} \quad < \frac{1}{60{,}000}$$

Example 2.35

Sally Clark. Sometimes coincidences are not happy events like lottery wins. In 1999, a British jury convicted Sally Clark of murdering her two children who had died suddenly at the ages of 11 and 8 weeks respectively of sudden infant death syndrome or "cot deaths." There was no physical or other evidence of a murder, nor was there a motive. Most likely, the jury was convinced by a pediatrician who said that a baby had a probability of roughly 1/8,500 of dying a cot death, so having two children die this way had probability roughly 1/73,000,000.

There are two problems with the computation: (i) Many families have children who die this way, so the first factor of 1/8,500 should be dropped. (ii) Two cot deaths in the same family are not independent events; once one occurs the second child faces an increased risk of about 1/100 of dying this way. Thus the probability that it would happen again without foul play is 1/100. If this number had been presented to the jury, Sally probably would not have had to spend 3 years in jail before the verdict was overturned.

2.4 Card games and other urn problems

A number of problems in probability have the following form.

Example 2.36 Suppose we pick 4 balls out of an urn with 12 red balls and 8 black balls. What is the probability of $B =$ "we get two balls of each color"?

Almost by definition, there are

$$C_{20,4} = \frac{20 \cdot 19 \cdot 18 \cdot 17}{1 \cdot 2 \cdot 3 \cdot 4} = 5 \cdot 19 \cdot 3 \cdot 17 = 4{,}845$$

ways of picking 4 balls out of the 20. To count the number of outcomes in B, we note that there are $C_{12,2}$ ways to choose the red balls and $C_{8,2}$ ways to choose the black balls, so the multiplication rule implies

$$|B| = C_{12,2}C_{8,2} = \frac{12 \cdot 11}{1 \cdot 2} \cdot \frac{8 \cdot 7}{1 \cdot 2} = 6 \cdot 11 \cdot 4 \cdot 7 = 1{,}848$$

It follows that $P(B) = 1{,}848/4{,}845 = 0.3814$.

We now consider two gambling games in which numbered balls are picked out of urns.

Example 2.37 **New York State lottery.** As mentioned in Section 2.1, if there are 59 numbered balls and 6 are picked, the number of outcomes is

$$C_{59,6} = 45{,}057{,}474$$

In the lottery you do win some money if at least three of your six numbers are chosen. The problem is to compute the probability of winning these other prizes.

Five out of six has probability

$$\frac{C_{6,5}C_{53,1}}{C_{59,6}} = \frac{6 \cdot 53}{C_{59,6}} = \frac{1}{141{,}690}$$

Four out of six has probability

$$\frac{C_{6,4}C_{53,2}}{C_{59,6}} = \frac{15 \cdot 1{,}378}{C_{59,6}} = \frac{1}{2{,}180}$$

Three out of six has probability

$$\frac{C_{6,3}C_{53,3}}{C_{59,6}} = \frac{20 \cdot 23{,}426}{C_{59,6}} = \frac{1}{96}$$

To add more prizes, a bonus number has been added to the card. You win if you match 5 out of 6 and also the bonus number that has probability

$$\frac{C_{6,5}C_{52,1}}{C_{59,6}}\frac{1}{52} = \frac{6}{C_{59,6}} = \frac{1}{7{,}509{,}579}$$

It is hard to compute the expected value because the rewards for 6 out of 6 and 5 out of 6 plus the bonus number depend on the number of weeks that there has been no winner, and all prizes with the exception of 3 out of 6 which always pays $1 depend on both the number of people who play and the number of people who win. However, one can get some idea of the expected value by noting that 54.7% of the money bet is returned in prizes, 32.9% to education, and 12.4% to various operating expenses.

The next table gives data for the number of winners and winning amounts for the month of January 2008.

Date	6	5+	5	4	3
1/2	10M	503,347	2,284	23	1
	0	0	15	1,601	36,719
1/5	11M	557,243	1,635	29	1
	0	1	25	1,502	36,203
1/9	12M	46,833	2,200	34	1
	1	0	16	1,128	26,319
1/12	3M	96,488	2,354	32	1
	0	1	16	1,299	30,689
1/16	4M	41,839	2,645	36	1
	0	0	12	951	24,147
1/19	5M	91,440	984	21	1
	0	0	37	1,909	42,927
1/23	6M	133,696	1,885	26	1
	0	0	17	1,319	31,131
1/26	7M	184,664	1,104	26	1
	0	1	34	1,582	37,648
1/30	8M	38,956	461	11	1
	2	1	60	2,700	45,386

As you can see from the table, the big prize starts at 3 million and increases by 1 million each week when there are no winners. The 5 out of 6 plus bonus number is about 40,000 times the number of weeks since the previous winner. One can get a pretty good idea of the number of people who played each week by multiplying the number of 3 out of 6 winners by 96 (or 100 which is easier).

Example 2.38

Keno. In this game the casino picks 20 balls out of 80 numbered balls. Before the draw you may, for example, pick 10 numbers and bet \$1. In this case you win \$1 if 4 of your numbers are chosen; \$2 for 5; \$20 for 6; \$105 for 7; \$500 for 8; \$5,000 for 9; and \$12,000 if all 10 are chosen. We want to compute the expected value of the bet.

The number of possible draws is astronomically large:

$$C_{80,20} = 3.5353 \times 10^{18}$$

The probability that k of your numbers are chosen is

$$p_k = \frac{C_{10,k} C_{70,20-k}}{C_{80,20}}$$

When $k = 0$, this is

$$\frac{C_{70,20}}{C_{80,20}} = \frac{70!60!}{80!50!} = \frac{60 \cdot 59 \cdot \cdots \cdot 51}{80 \cdot 79 \cdot \cdots \cdot 71} = 0.045791$$

To compute the other probabilities it is useful to note that for $1 \le m \le n$,

$$\begin{aligned} C_{n,m} = \frac{n!}{m!(n-m)!} &= \frac{n+1-m}{m} \cdot \frac{n!}{(m-1)!(n+1-m)!} \\ &= \frac{n+1-m}{m} \cdot C_{n,m-1} \end{aligned}$$

so we have

$$p_k = p_{k-1} \cdot \frac{11-k}{k} \cdot \frac{21-k}{50+k}$$

Writing w_k for the winning when k of our numbers are drawn, using this recursion and the result for p_0 gives

k	p_k	w_k	$w_k p_k$
0	0.045791		0
1	0.179571		0
2	0.295257		0
3	0.267402		0
4	0.147319	1	0.147319
5	0.051428	2	0.102855
6	0.011479	20	0.229588
7	0.001611	105	0.169701
8	0.000135	500	0.067710
9	6.12×10^{-6}	5,000	0.030603
10	1.12×10^{-7}	12,000	0.001347
4–10	0.2120		0.7486

Thus, we win something about 21.2% of the time and our average winning is a little less than 75 cents, a typical expected value for Keno bets. The last column shows the contribution of the different payoffs to the expected value.

Example 2.39 **Bridge.** In the game of bridge there are four players called North, West, South, and East according to their positions at the table. Each player gets 13 cards. The game is somewhat complicated, so we will content ourselves to analyze one situation that is important in the play of the game. Suppose that North and South have a total of 8 hearts. What is the probability that West will have 3 and East will have 2?

Even though this is not how the cards are usually dealt, we can imagine that West randomly draws 13 cards from the 26 that remain. This can be done in

$$C_{26,13} = \frac{26!}{13!\,13!} = 10{,}400{,}600 \text{ ways}$$

North and South have 8 hearts and 18 nonhearts, so in the 26 that remain there are $13 - 8 = 5$ hearts and $39 - 18 = 21$ nonhearts. To construct a hand for West with 3 hearts and 10 nonhearts we must pick 3 of the 5 hearts, which can be done in $C_{5,3}$ ways, and 10 of the 21 nonhearts in $C_{21,10}$. The multiplication rule then implies that the number of outcomes for West with 3 hearts is $C_{5,3} \cdot C_{21,10}$ and the probability of interest is

$$\frac{C_{5,3} \cdot C_{21,10}}{C_{26,13}} = 0.339$$

Multiplying by 2 gives the probability that one player will have 3 cards and the other 2, something called a 3–2 split. Repeating the reasoning gives that an $i - j$

split ($i + j = 5$) has probability

$$2 \cdot \frac{C_{5,i} \cdot C_{21,13-i}}{C_{26,13}}$$

This formula tells us that the probabilities are

3–2	0.678
4–1	0.282
5–0	0.039

Thus while a 3–2 split is the most common, one should not ignore the possibility of a 4–1 split. Similar calculations show that if North and South have 9 hearts then the probabilities are

2–2	0.406
3–1	0.497
4–0	0.095

In this case the uneven 3–1 split is more common than the 2–2 split since it can occur two ways; that is, West might have 3 or 1.

Example 2.40 **Disputed elections.** In a close election in a small town, 2,656 people voted for candidate A compared to 2,594 who voted for candidate B, a margin of victory of 62 votes. An investigation of the election, instigated no doubt by the loser, found that 136 of the people who voted in the election should not have. Since this is more than the margin of victory, should the election results be thrown out even though there was no evidence of fraud on the part of the winner's supporters?

Like many problems that come from the real world (a court case *De Martini v. Power*), this one is not precisely formulated. To turn this into a probability problem we suppose that all the votes were equally likely to be one of the 136 erroneously cast and we investigate what happens when we remove 136 balls from an urn with 2,656 white balls and 2,594 black balls. Now the probability of removing exactly m white and $136 - m$ black balls is

$$\frac{C_{2,656,m} C_{2,594,136-m}}{C_{5,250,136}}$$

In order to reverse the outcome of the election, we must have

$$2{,}656 - m \leq 2{,}594 - (136 - m) \quad \text{or} \quad m \geq 99$$

With the help of a short computer program we can sum the probability above from $m = 99$ to 136 to conclude that the probability of the removal of 136 randomly chosen votes reversing the election is 7.492×10^{-8}. This computation supports the Court of Appeals decision to overturn a lower court ruling that voided the election in this case.

Exercise. In election considered in *Ipolito v. Power*, the winning margin was 1,422 to 1,405 but 101 votes had to be thrown out. The judge rules that "it does not strain the probabilities to assume a likelihood that the questioned votes produced or could produce a change in the result." Do you agree with this assessment? We return to this question in Example 6.22.

Example 2.41

Quality control. A shipment of 50 precision parts including 4 that are defective is sent to an assembly plant. The quality control division selects 10 at random for testing and rejects the entire shipment if 1 or more are found defective. What is the probability that this shipment passes inspection?

There are $C_{50,10}$ ways of choosing the test sample and $C_{46,10}$ ways of choosing all good parts, so the probability is

$$\frac{C_{46,10}}{C_{50,10}} = \frac{46!/36!10!}{50!/40!10!} = \frac{46 \cdot 45 \cdots 37}{50 \cdot 49 \cdots 41}$$
$$= \frac{40 \cdot 39 \cdot 38 \cdot 37}{50 \cdot 49 \cdot 48 \cdot 47} = 0.396$$

Using almost identical calculations a company can decide on how many bad units they will allow in a shipment and design a testing program with a given probability of success.

Example 2.42

Capture–recapture experiments. An ecology graduate student goes to a pond and captures $k = 60$ beetles, marks each with a dot of paint, and then releases them. A few days later she goes back and captures another sample of $r = 50$, finding $m = 12$ marked beetles and $r - m = 38$ unmarked. What is her best guess about the size of the population of beetles?

To turn this into a precisely formulated problem, we suppose that no beetles enter or leave the population between the two visits. With this assumption, if there were N beetles in the pond, then the probability of getting m marked and $r - m$ unmarked in a sample of r would be

$$p_N = \frac{C_{k,m}\, C_{N-k,r-m}}{C_{N,r}}$$

To estimate the population we pick N to maximize p_N, the so-called **maximum likelihood estimate.** To find the maximizing N, we note that

$$C_{j-1,i} = \frac{(j-1)!}{(j-i-1)!i!} \quad \text{so} \quad C_{j,i} = \frac{j!}{(j-i)!i!} = \frac{jC_{j-1,i}}{(j-i)}$$

and it follows that

$$p_N = p_{N-1} \cdot \frac{N-k}{N-k-(r-m)} \cdot \frac{N-r}{N}$$

Now $p_N/p_{N-1} \geq 1$ if and only if

$$(N-k)(N-r) \geq N(N-k-r+m)$$

that is,

$$N^2 - kN - rN + kr \geq N^2 - kN - rN + mN$$

or equivalently if $N \leq kr/m$. Thus the value of N that maximizes the probability p_N is the largest integer $\leq kr/m$. This choice is reasonable since when $N = kr/m$, the proportion of marked beetles in the population, k/N, equals the proportion of marked beetles in the sample, m/r. Plugging in the numbers from our example, $kr/m = (60 \cdot 50)/12 = 250$, so the probability is maximized when $N = 250$.

2.5 Probabilities of unions, Joe DiMaggio

In Section 1.1, we learned that $P(A \cup B) = P(A) + P(B) - P(A \cap B)$. In this section we extend this formula to $n > 2$ events. We begin with $n = 3$ events:

$$\begin{aligned} P(A \cup B \cup C) = P(A) + P(B) + P(C) \\ - P(A \cap B) - P(A \cap C) - P(B \cap C) \\ + P(A \cap B \cap C) \end{aligned} \tag{2.15}$$

Proof. As in the proof of the formula for two events, we have to convince ourselves that the net number of times each part of $A \cup B \cup C$ is counted is 1. To do this, we make a table that identifies the areas counted by each term and note that the net number of pluses in each row is 1:

	A	B	C	$A \cap B$	$A \cap C$	$B \cap C$	$A \cap B \cap C$
$A \cap B \cap C$	+	+	+	−	−	−	+
$A \cap B \cap C^c$	+	+		−			
$A \cap B^c \cap C$	+		+		−		
$A^c \cap B \cap C$		+	+			−	
$A \cap B^c \cap C^c$	+						
$A^c \cap B \cap C^c$		+					
$A^c \cap B^c \cap C$			+				

Example 2.43

Suppose we roll three dice. What is the probability that we get at least one 6?

Let A_i = "we get a 6 on the ith die." Clearly,

$$P(A_1) = P(A_2) = P(A_3) = 1/6$$
$$P(A_1 \cap A_2) = P(A_1 \cap A_3) = P(A_2 \cap A_3) = 1/36$$
$$P(A_1 \cap A_2 \cap A_3) = 1/216$$

So plugging into (2.15) gives

$$P(A_1 \cup A_2 \cup A_3) = 3 \cdot \frac{1}{6} - 3 \cdot \frac{1}{36} + \frac{1}{216} = \frac{108 - 18 + 1}{216} = \frac{91}{216}$$

To check this answer, we note that $(A_1 \cup A_2 \cup A_3)^c$ = "no 6" = $A_1^c \cap A_2^c \cap A_3^c$ and $|A_1^c \cap A_2^c \cap A_3^c| = 5 \cdot 5 \cdot 5 = 125$ since there are five "non-6's" that we can get on each roll. Since there are $6^3 = 216$ outcomes for rolling three dice, it follows that $P(A_1^c \cap A_2^c \cap A_3^c) = 125/216$ and $P(A_1 \cup A_2 \cup A_3) = 1 - P(A_1^c \cap A_2^c \cap A_3^c) = 91/216$.

The same reasoning applies to sets.

Example 2.44

In a freshman dorm, 60 students read the *Cornell Daily Sun*, 40 read the *New York Times*, and 30 read the *Ithaca Journal*. 20 read the *Cornell Daily Sun* and the *New York Times*, 15 read the *Cornell Daily Sun* and the *Ithaca Journal*, 10 read the *New York Times* and the *Ithaca Journal*, and 5 read all three. How many read at least one newspaper?

Using our formula the answer is

$$60 + 40 + 30 - 20 - 15 - 10 + 5 = 90$$

To check this we can draw picture using D, N, and I for the three newspapers.

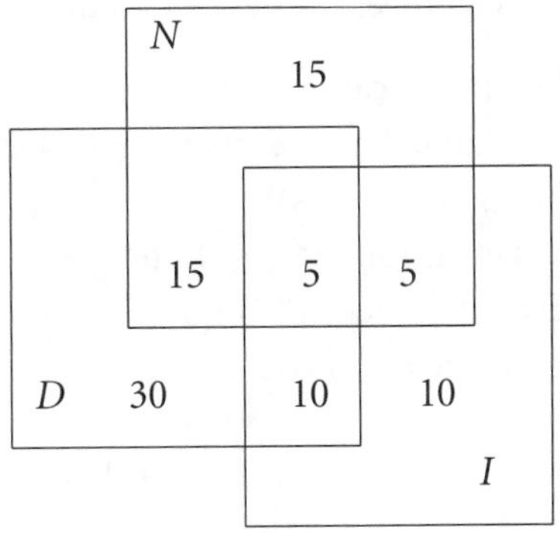

To figure out the number of students in each category we work out from the middle. $D \cap N \cap I$ has 5 students and $D \cap N$ has 20, so $D \cap N \cap I^c$ has 15. In the same way we compute that $D \cap N^c \cap I$ has $15 - 5 = 10$ students and

$D^c \cap N \cap I$ has $10 - 5 = 5$ students. Having found that 30 of students in D read at least one other newspaper, the number who read only D is $60 - 30 = 30$. In a similar way, we compute that there are $40 - 25 = 15$ students who read only N and $30 - 20 = 10$ students who read only I. Adding up the numbers in the seven regions gives a total of 90, as we found before.

2.5.1 Inclusion–exclusion formula

Formula (2.15) generalizes to n events:

$$P\left(\bigcup_{i=1}^{n} A_i\right) = \sum_{i=1}^{n} P(A_i) - \sum_{i<j} P(A_i \cap A_j) + \sum_{i<j<k} P(A_i \cap A_j \cap A_k) + \cdots + (-1)^{n+1} P(A_1 \cap \cdots \cap A_n) \tag{2.16}$$

In words, we take all possible intersections of $1, 2, \ldots, n$ events and the signs of the sums alternate.

Proof. A point that is in exactly k sets is counted k times by the first sum, $C_{k,2}$ times by the second, $C_{k,3}$ times by the third, and so on until it is counted $C_{k,k} = 1$ time by the kth term. The net result is

$$C_{k,1} - C_{k,2} + C_{k,3} + \cdots + (-1)^{k+1}1$$

To show that this adds up to 1, we recall the binomial theorem

$$(a+b)^k = a^k + C_{k,1}a^{k-1}b + C_{k,2}a^{k-2}b^2 + \cdots + b^k$$

Setting $a = 1$ and $b = -1$, we have

$$0 = 1 - C_{k,1} + C_{k,2} - C_{k,3} - \cdots - (-1)^{k+1}$$

which proves the desired result. □

Example 2.45 You pick 7 cards out of deck of 52. What is the probability that you have a three of a kind, that is, exactly three cards of some denomination (for example, three kings or three 7's)?

Let A_i for $1 \le i \le 13$ be the event you have three cards of type i where 1 is ace, 11 is jack, 12 is queen, and 13 is king. It is impossible for three of these events to occur so

$$P\left(\bigcup_{i=1}^{13} A_i\right) = 13P(A_1) - C_{13,2}P(A_1 \cap A_2)$$

A_1 can occur in $C_{4,3}C_{48,4} = 778{,}320$ ways and $A_1 \cap A_2$ can occur in $(C_{4,3})^2 \cdot 44 = 704$ ways so the answer is

$$\frac{13 \cdot 778{,}320 - 78 \cdot 704}{C_{52,7}} = \frac{10{,}118{,}160 - 54{,}912}{133{,}784{,}560} = 0.075219$$

Notice that the first term gives most of the answer and the second is only a small correction to account for the rare event of having two sets of three of a kind.

Example 2.46 Suppose we roll a die 15 times. What is the probability that we do not see each of the 6 numbers at least once?

Let A_i be the event that we never see i. $P(A_i) = 5^{15}/6^{15}$ since there are 6^{15} outcomes in all but only 5^{15} that contain no i's. $5^{15}/6^{15} = 0.064905$, so

$$\sum_{i=1}^{6} P(A_i) = 6(0.064905) = 0.389433$$

Turning to the second term, we note that for any $i < j$, we have $P(A_i \cap A_j) = 4^{15}/6^{15} = 0.002284$ and there are $C_{6,2} = (6 \cdot 5)/2 = 15$ choices for $i < j$, so

$$\sum_{i<j} P(A_i \cap A_j) = 15(0.002284) = 0.03426$$

For the third term, we note that for any $i < j < k$, we have $P(A_i \cap A_j \cap A_k) = 3^{15}/6^{15} = 3.05 \times 10^{-5}$ and there are $C_{6,3} = (6 \cdot 5 \cdot 4)/3! = 20$ choices for $i < j < k$, so

$$\sum_{i<j<k} P(A_i \cap A_j \cap A_k) = 20(3.05 \times 10^{-5}) = 0.00061$$

At this point the pattern should be clear:

$$\begin{aligned} &C_{6,1}(5/6)^{15} - C_{6,2}(4/6)^{15} + C_{6,3}(3/6)^{15} - C_{6,4}(2/6)^{15} + C_{6,5}(1/6)^{15} \\ &= 0.389433 - 0.03426 + 6.1 \times 10^{-4} - 1.045 \times 10^{-6} + 1.276 \times 10^{-11} \\ &= 0.355787 \end{aligned}$$

2.5.2 Bonferroni inequalities

In brief, if you stop the inclusion–exclusion formula with a + term you get an upper bound; if you stop with a − term you get a lower bound.

$$P\left(\bigcup_{i=1}^{n} A_i\right) \le \sum_{i=1}^{n} P(A_i) \tag{2.17}$$

$$\ge \sum_{i=1}^{n} P(A_i) - \sum_{i<j} P(A_i \cap A_j) \tag{2.18}$$

$$\le \sum_{i=1}^{n} P(A_i) - \sum_{i<j} P(A_i \cap A_j) + \sum_{i<j<k} P(A_i \cap A_j \cap A_k) \tag{2.19}$$

To explain the usefulness of these inequalities, we note that in the previous example they imply that the probability of interest

$$\begin{aligned}
&\leq 0.389433 \\
&\geq 0.389433 - 0.03426 = 0.355177 \\
&\leq 0.389433 - 0.03426 + 6.1 \times 10^{-4} = 0.355738
\end{aligned}$$

so we have a very accurate result after three terms.

Proof. The first inequality is obvious since the right-hand side counts each outcome in $\cup_{i=1}^{n} A_i$ at least once. To prove the second, consider an outcome that is in exactly k sets. If $k = 1$, the first term will count it once and the second not at all. If $k = 2$, the first term counts it twice and the second once, with a net total of 1. If $k \geq 3$, the first term counts it k times and the second $C_{k,2} = k(k-1)/2 > k$ times so the net number of countings is < 0.

The third formula is similar.

In k sets	Counted
1	$1 - 0 + 0 = 1$
2	$2 - 1 + 0 = 1$
3	$3 - 3 + 1 = 1$

When ≥ 4, the number of countings is

$$C_{k,1} - C_{k,2} + C_{k,3} > k - \frac{k(k-1)}{2} + (k-1)(k-2) \geq 0 \qquad \square$$

Example 2.47 **The streak.** In the summer of 1941, Joe DiMaggio achieved what many people consider the greatest record in sports, in which he had at least one hit in each of 56 games. What is the probability of this event?

A useful trick. Suppose for the moment that we know the probability p that Joe DiMaggio gets a hit in one game and that successive games are independent. Assuming a 154-game season, we could let A_i be the probability that a player got hits in games $i+1, \ldots, i+56$ for $0 \leq i \leq 98$. Using (2.17) it follows that the probability of the streak is

$$\leq 99\, p^{56}$$

As we will see in a minute this overestimates the actual answer by a factor of $1/(1-p)$. The problem is that if A_i occurs, it becomes much easier for A_{i+1}, A_{i-1}, and other "nearby" events to occur. To avoid this problem, we will let B_i be the event the player gets no hit in game i but has hits in games $i+1, i+2, \ldots, i+56$, where $1 \leq i \leq 98$. Ignoring the probability of having hits in games 1, 2, ..., 56, the event of interest $S = \cup_{i=1}^{98} B_i$, so

$$P(S) \leq q \equiv 98p^{56}(1-p)$$

To compute the second bound we begin by noting $B_i \cap B_j = \emptyset$ if $i < j \leq i+56$ since B_i requires a hit in game j, while B_j requires no hit. If $56 + i < j \leq 98$, then $P(B_i \cap B_j) = P(B_i)P(B_j)$. To simplify the arithmetic we note that in either case $P(B_i \cap B_j) \leq P(B_i)P(B_j)$, so

$$\sum_{1 \leq i < j \leq 98} P(B_i \cap B_j) \leq C_{98,2}\, p^{112}(1-p)^2 \leq \frac{q^2}{2}$$

This is the number we have to subtract from the upper bound to get the lower bound, so we have

$$q \geq P(S) \geq q - \frac{q^2}{2} \tag{2.20}$$

Since q will end up being very small, the ratio of the two bounds is $1 - (q/2) \approx 1$.

To compute the probability p that Joe DiMaggio gets a hit in one game, we will introduce two somewhat questionable assumptions: (i) A player gets exactly four at bats per game (during the streak, DiMaggio averaged 3.98 at bats per game) and (ii) the outcomes of different at bats are independent with the probability of a hit being 0.325, Joe DiMaggio's lifetime batting average. From assumptions (i) and (ii) it follows that the probability

$$p = 1 - (0.675)^4 = 0.7924$$

and using (2.20) we have

$$P(S) \approx q = 98(0.7924)^{56}(0.2076) = 4.46 \times 10^{-5}$$

To interpret our result, note that the probability in (2.20) is roughly 1/22,000, so even if there were 220 players with 0.325 batting averages, it would take 100 years for this to occur again.

Example 2.48

A less famous streak. *Sports Illustrated* reports that a high school football team in Bloomington, Indiana, lost 21 straight pregame coin flips before finally winning one. Taking into account the fact that there are approximately 15,000 high school and college football teams, is this really surprising?

We will first compute the probability that this happens to one team some time in the decade 1995–2004, assuming that the team plays 10 games per year. Taking a lesson from the previous example, we let B_i be the event that the team won the coin flip in game i but lost it in games $i+1, \ldots, i+21$. Using the reasoning that led to (2.20),

$$P(S) \approx 79(1/2)^{22} = 1.883 \times 10^{-5}$$

What we have computed is the probability that one particular team will have this type of bad luck some time in the last decade. The probability that none of the 15,000 teams will do this is

$$(1 - 79(0.5)^{22})^{15,000} = 0.7539$$

that is, with probability 0.2461 some team will have this happen to them. As a check on the last calculation, note that (2.17) gives an upper bound of

$$15,000 \times 1.883 \times 10^{-5} = 0.2825$$

2.6 Blackjack

In this book we analyze craps and roulette, casino games where the player has a substantial disadvantage. In the case of blackjack, a little strategy, which we explain in this section, can make the game almost even. To begin we describe the rules and the betting.

In the game of blackjack, a king, queen, or jack counts 10, an ace counts 1 or 11, and the other cards count the numbers that are shown on them (for example, a 5 counts 5). The object of the game is to get as close to 21 as you can without going over. You start with 2 cards and draw cards out of the deck until either you are happy with your total or you go over 21, in which case you "bust."

If your initial two cards total 21, this is a blackjack, and if the dealer does not have one, you win 1.5 times your original bet. If you bust then you immediately lose your bet. This is the main source of the casino advantage since if the dealer busts later you have already lost. If you stop with 21 or less and the dealer busts, you win. If you and the dealer both end with 21 or less then the one with higher hand wins. In the case of a tie no money changes hands.

In casino blackjack the dealer plays by a simple rule: He draws a card if his total is ≤ 16, otherwise he stops. The first step in analyzing blackjack is to compute the probability that the dealer's ending total is k when he has a total of j. To deal with the complication that an ace can count as 1 or 11, we introduce $b(j, k) =$ the probability that the dealer's ending total is k when he has a total of j including one ace that is being counted as 11. Such hands are called **soft** because even if you draw a 10, you will not bust. We define $a(j, k) =$ the probability that the dealer's ending total is k when he has a hard total of j, that is, a hand in which any ace is counted as 1.

We start by observing that $a(j, j) = b(j, j) = 1$ when $j \geq 17$ and then start with 16 and work down. Let $p_i = 1/13$ for $1 \leq i < 9$ and $p_{10} = 4/13$. If $11 \leq j \leq 16$, then a new ace must count as 1, so

$$a(j, k) = p_1 a(j+1, k) + \sum_{m=2}^{10} p_m a(j+m, k)$$

When $2 \leq j \leq 10$, a new ace counts as 11 and produces a soft hand:

$$a(j, k) = p_1 b(j+11, k) + \sum_{m=2}^{10} p_m a(j+m, k)$$

For soft hands, an ace counts as 11, so there are no soft hands with totals of less than 12. If the card we draw takes us over 21 then we have to change the ace from counting 11 to counting 1, producing a hard hand, so

$$b(j, k) = p_1 b(j+1, k) + \sum_{m=2}^{21-j} p_m b(j+m, k) + \sum_{m=22-j}^{10} p_m a(j+m-10, k)$$

When $j = 11$, the second sum runs from 11 to 10 and is considered to be 0.

The last three formulas are too complicated to work with by hand but are easy to manipulate using a computer. The next table gives the probabilities of the various results for the dealer conditional on the value of his first card. We have broken things down this way because when blackjack is played in a casino, we can see one of the dealer's two cards.

	17	18	19	20	21	Bust
2	0.13981	0.13491	0.12966	0.12403	0.11799	**0.35361**
3	0.13503	0.13048	0.12558	0.12033	0.11470	**0.37387**
4	0.13049	0.12594	0.12139	0.11648	0.11123	**0.39447**
5	0.12225	0.12225	0.11770	0.11315	0.10825	**0.41640**
6	0.16544	0.10627	0.10627	0.10171	0.09716	**0.42315**
7	**0.36857**	0.13780	0.07863	0.07863	0.07407	0.26231
8	0.12857	**0.35934**	0.12857	0.06939	0.06939	0.24474
9	0.12000	0.12000	**0.35076**	0.12000	0.06082	0.22843
10	0.11142	0.11142	0.11142	**0.34219**	0.11142	0.21211
Ace	0.13079	0.13079	0.13079	0.13079	**0.36156**	0.11529

You should note that when the dealer's upcard is 2, 3, 4, 5, or 6, her most likely outcome is to bust, but when her first card is $k = 7, 8, 9, 10$, or ace $= 11$, her most likely total is $10 + k$. To make this clear we have given the most likely probabilities in boldface.

The analysis of the player's options is even more complicated than that of the dealer, so we will not attempt it here. The first analysis was performed in the mid-1950s (see Baldwin et al. in *Journal of the American Statistical Association*, Vol. 51, pp. 429–439) and has been redone by a number of other people since that time. To describe the optimal strategy in a few words we use "stand on n" as short for "take a card if your total is $< n$ but not if it is $\geq n$."

Hard hands

Stand on 17 if the dealer shows 7, 8, 9, 10, or A.
Stand on 12 if the dealer shows 2, 3, 4, 5, or 6.
Exception: Draw to 12 if the dealer shows 2 or 3.

Soft hands

Stand on 18.
Exception: Draw to 18 if the dealer has 9 or 10.

To help remember the rules for hard hands, observe that with two exceptions the strategy there is a combination of "mimic the dealer" and "never bust" (that is, "only take a card if you have 11 or less"), and it is exactly what we would do if the dealer's downcard was a 10. If her upcard is 7, 8, 9, 10, or A, then we must get to 17 to have a chance of winning. If her upcard is 2, 3, 4, 5, or 6 then we don't draw and hope that she busts.

Using these rules, the probability that you will win is about 0.49, close enough to even if you are only looking for an evening's entertainment. You can reduce the house edge even further by learning about "doubling down" and splitting pairs.

Doubling down. In this move you turn up your two cards, double your bet, and ask for one card to be dealt down to you. You are not allowed to ask for a second card if you don't like the first one. Double down

- If your total is 11
- If your total is 10 and the dealer's upcard is 9 or less
- If your total is 9 and the dealer's upcard is 3 through 6
- If you have A and 2 through 7 and the dealer's upcard is 4, 5, or 6

Again the doubling down rules can be explained by assuming that we are going to get a 10. Some Nevada casinos allow doubling down only on 11 or 10.

Splitting pairs. If you have a pair you can split them, an extra card is dealt to each one, you place another bet on the table, so there is one on each hand, and then play two hands separately.

- Always split A's or 8's.
- Never split 4's, 5's, or 10's.

- Split 2's, 3's, 6's, and 7's when the dealer's upcard is 3 through 7.
- Split 9's when the dealer's upcard is 2 through 9, but not 7.

The reason for splitting aces should be obvious. It is such a good play that it has on occasion been forbidden. Some casino rules do not allow further drawing after aces are split, and if a 10 lands on the ace, it is not a blackjack. To see why 8's are singled out for splitting, note that $8 + 8 = 16$, which wins only if the dealer busts, while an 8 paired with a 10 produces an 18.

Counting cards. Edward Thorp's book *Beat the Dealer*, which astonished the world in 1962 by demonstrating that by "counting cards" (that is, by keeping track of the difference between the numbers of cards you have seen that count 10 and those that count 2 through 6) and by adjusting your betting you can make money from blackjack. Before the reader plans a trip to Las Vegas or Atlantic City, we would like to point out that playing this strategy requires hardwork, that making money with it requires a lot of capital, and that casinos are allowed to ask you to leave if they think you are playing it. The book *Bringing Down the House* gives an entertaining account of MIT students using the strategy to win money at blackjack.

2.7 Exercises

Permutations and combinations

1. How many possible batting orders are there for nine baseball players?

2. A tire manufacturer wants to test four different types of tires on three different types of roads at five different speeds. How many tests are required?

3. 16 horses race in the Kentucky Derby. How many possible results are there for win, place, and show (first, second, and third)?

4. A school gives awards in five subjects to a class of 30 students but no one is allowed to win more than one award. How many outcomes are possible?

5. A tourist wants to visit six of America's ten largest cities. In how many ways can she do this if the order of her visits is (a) important or (b) not important?

6. Five businessmen meet at a convention. How many handshakes are exchanged if each shakes hands with all the others?

7. A commercial for Glade Plug-ins says that by inserting 2 of a choice of 11 scents into the device, you can make more than 50 combinations. If we exclude the boring choice of two of the same scent, how many possibilities are there?

8. In a class of 19 students, 7 will get A's. In how many ways can this set of students be chosen?

9. (a) How many license plates are possible if the first three places are occupied by letters and the last three by numbers? (b) Assuming all combinations are equally likely, what is the probability that the three letters and the three numbers are different?

10. How many four-letter "words" can you make if no letter is used twice and each word must contain at least one vowel (A, E, I, O, or U)?

11. Assuming all phone numbers are equally likely, what is the probability that all the numbers in a seven-digit phone number are different?

12. A domino is an ordered pair (m, n) with $0 \le m \le n \le 6$. How many dominoes are in a set if there is only one of each?

13. A person has 12 friends and will invite 7 to a party. (a) How many choices are possible if Al and Bob are feuding and will not both go to the party? (b) How many choices are possible if Al and Betty insist that they both go or neither one goes?

14. A basketball team has 5 players more than 6 feet tall and 6 who are less than 6 feet. How many ways can they have their picture taken if the 5 taller players stand in a row behind the 6 shorter players who are sitting on a row of chairs?

15. The Duke basketball team has 10 women who can play guard and 12 tall women who can play the other three positions. At the start of the game, the coach gives the referee a starting lineup that lists who will play left guard, right guard, left forward, center, and right forward. In how many ways can this be done?

16. Six students, three boys and three girls, lineup in a random order for a photograph. What is the probability that the boys and girls alternate?

17. Seven people sit at a round table. How many ways can this be done if Mr. Jones and Miss Smith (a) must sit next to each other and (b) must not sit next to each other? (Two seating patterns that differ only by a rotation of the table are considered the same.)

18. How many ways can four rooks be put on a chessboard so that no rook can capture any other rook? Or, what is the same: How many ways can 8 markers be placed on an 8×8 grid of squares so that there is at most one in each row or column?

19. A BINGO card is a 5×5 grid. The center square is a free space and has no number. The first column is filled with five distinct numbers from 1 to 15, the second with five numbers from 16 to 30, the middle column with four numbers from 31 to 45, the fourth with five numbers from 46 to 60, and the fifth with five numbers from 61 to 75. Since the object of the game is to get five in a row horizontally, vertically, or diagonally, the order is important. How many BINGO cards are there?

20. Continuing with the setup from the previous problem, in BINGO numbers are drawn from 1 to 75 without replacement. When a number is called you put a marker on that square. If you have five in a row horizontally, vertically, or diagonally, you have a BINGO. What is the probability you will have a BINGO after (a) four numbers are called? (b) After five?

Multinomial counting problems

21. How many different ways can the letters in the following words be arranged: (a) money, (b) banana, (c) statistics, (d) Mississippi?

22. 12 different toys are to be divided among 3 children so that each one gets 4 toys. How many ways can this be done?

23. A club with 50 members is going to form two committees, one with 8 members and the other with 7. How many ways can this be done (a) if the committees must be disjoint? (b) If they can overlap?

24. If seven dice are rolled, what is the probability that each of the six numbers will appear at least once?

25. How many ways can 5 history books, 3 math books, and 4 novels be arranged on a shelf if the books of each type must be together?

26. Suppose three runners from team A and three runners from team B have a race. If all six runners have equal ability, what is the probability that the three runners from team A will finish first, second, and fourth?

27. Four men and four women are shipwrecked on a tropical island. How many ways can they (a) form four male–female couples, (b) get married if we keep track of the order in which the weddings occur, (c) divide themselves into four unnumbered pairs, (d) split up into four groups of two to search the North, East, West, and South shores of the island, (e) walk single file up the ramp to the ship when they are rescued, (f) take a picture to remember their ordeal if all eight stand in a line but each man stands next to his wife?

Binomial and multinomial distributions

28. A die is rolled 8 times. What is the probability that we will get exactly two 3's?

29. Mary knows the answers to 20 of the 25 multiple-choice questions on the Psychology 101 exam, but she has skipped several of the lectures; she must take random guesses for the other five. Assuming each question has four answers, what is the probability that she will get exactly 3 of the last 5 questions right?

30. In 1997, 10.8% of female smokers smoked cigars. In a sample of size 10 female smokers, what is the probability that (a) exactly 2 of the women smoke cigars? (b) At most 1 smokes cigars?

31. A 1994 report revealed that 32.6% of U.S. births were to unmarried women. A parenting magazine selected 30 women who gave birth in 1994 at random. (a) What is the probability that exactly 10 of the women were unmarried? (b) Using your calculator determine the probability that in the sample at most 10 are unmarried.

32. 20% of all students are left-handed. A class of size 20 meets in a room with 5 left-handed and 18 right-handed chairs. Use your calculator to find the probability that each student will have a chair to match their needs.

33. David claims to be able to distinguish brand B beer from brand H, but Alice claims that he just guesses. They set up a taste test with 10 small glasses of beer. David wins if he gets 8 or more right. What is the probability that he will win (a) if he is just guessing? (b) If he gets the right answer with probability 0.9?

34. The following situation comes up the game of Yahtzee. We have three rolls of five dice and want to get three sixes or more. On each turn we reroll any dice that are not 6's. What is the probability that we succeed?

35. A baseball pitcher throws a strike with probability 0.5 and a ball with probability 0.5. He is facing a batter who never swings at a pitch. What is the probability that he strikes out, that is, gets three strikes before four balls?

36. A baseball player is said to "hit for the cycle" if he has a single, a double, a triple, and a home run all in one game. Suppose these four types of hits have probabilities 1/6, 1/20, 1/120, and 1/24. What is the probability of hitting for the cycle if he gets to bat (a) four times and (b) five times? (c) Using $P(\cup_i A_i) \le \sum_i P(A_i)$ shows that the answer to (b) is at most 5 times the answer to (a). What is the ratio of the two answers?

Poisson approximation

37. Compare the Poisson approximation with the exact binomial probabilities when (a) $n = 10$, $p = 0.1$, (b) $n = 20$, $p = 0.05$, and (c) $n = 40$, $p = 0.025$.

38. Use the Poisson approximation to compute the probability that you will roll at least one double 6 in 24 trials. How does this compare with the exact answer?

39. The probability of a three of a kind in poker is approximately 1/50. Use the Poisson approximation to compute the probability that you will get at least one three of a kind if you play 20 hands of poker.

40. Calls to a toll-free hotline service are made randomly at rate 2 per minute. The service has five operators, none of whom is currently busy. Use the Poisson distribution to estimate the probability that in the next minute there are < 5 calls.

41. In one of the New York state lottery games, a number is chosen at random between 0 and 999. Suppose you play this game 250 times. Use the Poisson approximation to estimate the probability that you will never win and compare this with the exact answer.

42. If you bet \$1 on number 13 at roulette (or on any other number) then you win \$35 if that number comes up, an event of probability 1/38, and you lose your dollar otherwise. Suppose you play 70 times. Use the Poisson approximation to estimate the probability that (a) you have won 0 times and lost \$70, and (b) you have won 1 time and lost \$34. (c) If you win 2 times you have won \$2. Combine the results of (a) and (b) to conclude that the probability that you will have won more money than you have lost is larger than 1/2.

43. In a particular Powerball drawing 210,850,582 tickets were sold. The chance of winning the lottery is 1 in 80,000,000. Use the Poisson approximation to estimate the probability that there is exactly one winner.

44. Suppose that the probability of a defect in a foot of magnetic tape is 0.002. Use the Poisson approximation to compute the probability that a 1,500-foot roll will have no defects.

45. Suppose 1% of a certain brand of Christmas lights is defective. Use the Poisson approximation to compute the probability that in a box of 25 there will be at most one defective bulb.

46. In February 2000, 2.8% of Colorado's labor force was unemployed. Calculate the probability that in a group of 50 workers exactly one is unemployed.

47. An insurance company insures 3,000 people, each of whom has a 1/1,000 chance of an accident in 1 year. Use the Poisson approximation to compute the probability that there will be at most 2 accidents.

48. Suppose that 1% of people in the population are more than 6 feet 3 inches tall. What is the chance that in a group of 200 people picked at random from the population at least four people will be more than 6 feet 3 inches tall.

49. In an average year in Mythica there are 8 fires. Last year there were 12 fires. How likely is it to have 12 or more fires just by chance?

50. An airline company sells 160 tickets for a plane with 150 seats, knowing that the probability that a passenger will not show up for the flight is 0.1. Use the Poisson approximation to compute the probability that they will have enough seats for all the passengers who show up.

51. Books from a certain publisher contain an average of 1 misprint per page. What is the probability that on at least one page in a 300-page book there are five misprints?

Urn problems

52. Two red cards and two black cards are lying face down on the table. You pick two cards and turn them over. What is the probability that the two cards are different colors?

53. Four people are chosen at random from 5 couples. What is the probability that two men and two women are selected?

54. You pick 5 cards out of a deck of 52. What is the probability that you get exactly 2 spades?

55. Seven students are chosen at random from a class with 17 boys and 13 girls. What is the probability that 4 boys and 3 girls are selected?

56. In a carton of 12 eggs, 2 are rotten. If we pick 4 eggs to make an omelet, what is the probability that we do not get a rotten egg?

57. An electronics store receives a shipment of 30 calculators of which 4 are defective. Six of these calculators are selected to be sent to a local high school. What is the probability that exactly one is defective?

58. A scrabble set contains 54 consonants, 44 vowels, and 2 blank tiles. Find the probability that your initial draw contains 5 consonants and 2 vowels.

59. (a) How many ways can we pick 4 students from a group of 40 to be on the math team? (b) Suppose there are 18 boys and 12 girls. What is the probability that the team will have 2 boys and 2 girls.

60. The following probability problem arose in a court case concerning possible discrimination against black nurses. 26 white nurses and 9 black nurses took an exam. All the white nurses passed but only 4 of the black nurses did. What is the probability that we would get this outcome if the five nurses who failed were chosen at random?

61. A closet contains 8 pairs of shoes. You pick out 5. What is the probability of (a) no pair, (b) exactly one pair, and (c) two pairs?

62. A drawer contains 10 black, 8 brown, and 6 blue socks. If we pick two socks at random, what is the probability that they match?

63. A dance class consists of 12 men and 10 women. Five men and five women are chosen and paired up to dance. In how many ways can this be done?

64. Suppose we pick 5 cards out of a deck of 52. What is the probability that we get at least one card of each suit?

65. A bridge hand in which there is no card higher than a 9 is called a *Yarborough* after the Earl who liked to bet at 1,000 to 1 that your bridge hand would have a card that was 10 or higher. What is the probability of a Yarborough when you draw 13 cards out of a deck of 52.

66. Two cards are a blackjack if one is an A and the other is a K, Q, J, or 10. (a) If you pick two cards out of a deck, what is the probability that you will get a blackjack? (b) Suppose you are playing blackjack against the dealer with a freshly shuffled deck. What is the probability that you or the dealer will get a blackjack?

67. A student studies 12 problems from which the professor will randomly choose 6 for a test. If the student can solve 9 of the problems, what is the probability she can solve at least 5 of the problems on the test?

68. A football team has 16 seniors, 12 juniors, 8 sophomores, and 4 freshmen. If we pick 5 players at random, what is the probability that we will get 2 seniors and 1 from each of the other 3 classes?

69. In a kindergarten class of 20 students, one child is picked each day to help serve the morning snack. What is the probability that in 1 week five different children are chosen?

70. An investor picks 3 stocks out of 10 recommended by his broker. Of these, 6 will show a profit in the next year. What is the probability that the investor will pick (a) 3, (b) 2, (c) 1, (d) 0 profitable stocks?

71. Four red cards (that is, hearts and diamonds) and four black cards are face down on the table. A psychic who claims to be able to locate the four black cards turns over 4 cards and gets 3 black cards and 1 red card. What is the probability that he would do this if he were guessing?

72. A town council considers the question of closing down an "adult" theater. The five men on the council all vote against this and the three women vote in favor. What is the probability that we would get this result (a) if the council members determined their votes by flipping a coin? (b) If we assigned the five "no" votes to council members chosen at random?

73. An urn contains white balls numbered 1 to 15 and black balls also numbered 1 to 15. Suppose you draw 4 balls. What is the probability that (a) no two have the same number? (b) You get exactly one pair with the same number? (c) You get two pair with the same numbers?

74. A town has four TV repairmen. In the first week of September four TV sets break and their owners call repairmen chosen at random. Find the probability that the number of repairmen who have jobs is 1, 2, 3, 4.

75. Compute the probabilities of the following poker hands when we roll five six-sided dice.

(a) Five of a kind	0.000771
(b) Four of a kind	0.019290
(c) A full house	0.038580
(d) Three of a kind	0.154320
(e) Two pair	0.231481
(f) One pair	0.462962
(g) No pair	0.092592

76. In seven-card stud you receive seven cards and use them to make the best poker hand you can. Ignoring the possibility of a straight or a flush the probability that the best hand you can make with your cards is

	Seven cards	Five cards
(a) Four of a kind	0.001680	0.000240
(b) A full house	0.025968	0.001441
(c) Three of a kind	0.049254	0.021128
(d) Two pair	0.240113	0.047539
(e) One pair	0.472839	0.422569
(f) No pair	0.210150	0.507082

Verify the probabilities for seven-card stud. Hint: For full house you need to consider hand patterns: 3–3–1 and 3–2–2 in addition to the more likely 3–2–1–1. For two pair you also have to consider the possibility of three pair.

Probabilities of unions

77. Six high school teams play each other in the Southern Tier division. Each team plays all the other teams once. What is the probability that some team has a perfect 5–0 season?

78. Suppose you draw 7 cards out of a deck of 52. What is the probability that you will have (a) exactly 5 cards of one suit? (b) At least 5 cards of one suit?

79. In a certain city 60% of the people subscribe to newspaper A, 50% to B, 40% to C, 30% to A and B, 20% to B and C, and 10% to A and C, but no one subscribes to all three. What percentage subscribe to (a) at least one newspaper and (b) exactly one newspaper?

80. Santa Claus has 45 drums, 50 cars, and 55 baseball bats in his sled. 15 boys will get a drum and a car, 20 a drum and a bat, 25 a bat and a car, and 5 will get three presents. (a) How many boys will receive presents? (b) How many boys will get just a drum?

81. Use the inclusion–exclusion formula to compute the probability that a randomly chosen number between 0000 and 9999 contains at least one 1. Check this by computing the probability that there is no 1.

82. Ten people call an electrician and ask him to come to their houses on randomly chosen days of the work week (Monday through Friday). What is the probability of $A =$ "he has at least one day with no jobs"?

83. We pick a number between 0 and 999, then a computer picks one at random from that range. Use (2.15) to compute the probability that at least two of our digits will match the computer's number. (Note: We include any leading zeros, so 017 and 057 have two matching digits.)

84. You pick 13 cards out of a deck of 52. What is the probability that you will not get a card from every suit?

85. You pick 13 cards out of a deck of 52. Let $A =$ "you have exactly 6 cards in at least one suit" and $B =$ "you have exactly 6 spades." The first Bonferroni inequality says that $P(A) \leq 4P(B)$. Compute $P(A)$ and $P(A)/P(B)$.

86. Use the first two Bonferroni inequalities to compute an upper and a lower bound on the probability that in a group of 60 people, at least 3 were born on the same day.

87. Suppose we roll two dice 6 times. Use the first three Bonferroni inequalities to compute bounds on the probability of $A =$ "we get at least one double 6." Compare the bounds with the exact answer $1 - (35/36)^6$.

88. Suppose we try 20 times for an event with probability 0.01. Use the first three Bonferroni inequalities to compute bounds on the probability of one success.

3

Conditional Probability

3.1 Definition

Suppose we are told that the event A with $P(A) > 0$ occurs. As explained in Section 1.3, then the sample space is reduced from Ω to A and by (1.6) the probability that B will occur given that A has occurred is

$$P(B|A) = \frac{P(B \cap A)}{P(A)} \tag{3.1}$$

Example 3.1 Suppose we roll two dice. Let $A =$ "the sum is 8" and $B =$ "the first die is 3." $A = \{(2,6), (3,5), (4,4), (5,3), (6,2)\}$, so $P(A) = 5/36$. $A \cap B = \{(3,5)\}$, so

$$P(B|A) = \frac{1/36}{5/36} = \frac{1}{5}$$

The same result holds if $B =$ "the first die is k" and $2 \le k \le 6$. Carrying this reasoning further, we see that given the outcome lies in A, all 5 possibilities have the same probability. This should not be surprising. The original probability is uniform over the 36 possibilities, so when we condition on the occurrence of A, its 5 outcomes are equally likely.

As the last example may have suggested, the mapping $B \to P(B|A)$ is a probability. That is, it is a way of assigning numbers to events that satisfies the axioms introduced in Chapter 1. To prove this, we note that

(i) $0 \le P(B|A) \le 1$ since $0 \le P(B \cap A) \le P(A)$.

(ii) $P(\Omega|A) = P(\Omega \cap A)/P(A) = 1$.

(iii) and (iv) If B_i are disjoint then $B_i \cap A$ are disjoint and $(\cup_i B_i) \cap A = \cup_i (B_i \cap A)$, so using the definition of conditional probability and parts (iii) and (iv) of the definition of probability we have

$$P(\cup_i B_i | A) = \frac{P(\cup_i (B_i \cap A))}{P(A)} = \frac{\sum_i P(B_i \cap A)}{P(A)} = \sum_i P(B_i|A)$$

From the last observation it follows that $P(\cdot|A)$ has the same properties that ordinary probabilities do; for example, if $C = B^c$,

$$P(C|A) = 1 - P(B|A) \tag{3.2}$$

Actually for this to hold, it is enough that B and C complement each other inside A; that is, $(B \cap C) \cap A = \emptyset$ and $(B \cup C) \supset A$.

Example 3.2

Alice and Bob are playing a gambling game. Each rolls one die and the person with the higher number wins. If they tie, they roll again. If Alice just won, what is the probability she rolled a 5?

Let $A =$ "Alice wins" and $R_i =$ "she rolls an i." If we write outcomes with Alice's roll first and Bob's second, the event A

$$\begin{array}{ccccc}
(2,1) & (3,1) & (4,1) & (5,1) & (6,1) \\
 & (3,2) & (4,2) & (5,2) & (6,2) \\
 & & (4,3) & (5,3) & (6,3) \\
 & & & (5,4) & (6,4) \\
 & & & & (6,5)
\end{array}$$

There are $1 + 2 + 3 + 4 + 5 = 15$ outcomes in A, and if we condition on A, they are all equally likely. $A \cap R_5$ has 4 outcomes, so $P(R_5|A) = 4/15$. In general, $P(R_i|A) = (i - 1)/15$ for $1 \leq i \leq 6$.

Example 3.3

A person picks 13 cards out of a deck of 52. Let $A_1 =$ "he has at least one ace," $H =$ "he has the ace of hearts," and $E_1 =$ "he receives exactly one ace." Find $P(E_1|A_1)$ and $P(E_1|H)$. Do you think these will be equal? If not then which one is larger?

Let $E_0 =$ "he has no ace."

$$p_0 = P(E_0) = \frac{C_{48,13}}{C_{52,13}} \qquad p_1 = P(E_1) = \frac{4C_{48,12}}{C_{52,13}}$$

Since $E_1 \subset A_1$ and $A_1 = E_0^c$,

$$P(E_1|A_1) = \frac{P(E_1)}{P(A_1)} = \frac{p_1}{1 - p_0}$$

Since $E_1 \cap H$ means you get the ace of hearts and no other ace:

$$P(E_1|H) = \frac{P(E_1 \cap H)}{P(H)} = \frac{C_{48,12}/C_{52,13}}{1/4} = p_1$$

To compare the probabilities we observe

$$P(E_1|A_1) = \frac{p_1}{1 - p_0} > p_1 = P(E_1|H)$$

Letting A_2 = "he has at least two aces" and using (3.2) we have

$$P(A_2|A_1) < P(A_2|H)$$

Intuitively, the event H is harder to achieve than A_1, so conditioning on it increases our chance of having other aces.

Multiplying the definition of conditional probability in (3.1) on each side by $P(A)$ gives the **multiplication rule**

$$P(A)P(B|A) = P(B \cap A) \tag{3.3}$$

Example 3.4

Suppose we draw 2 cards out of a deck of 52. What is the probability that both cards are spades?

Let A = "the first card is a spade" and B = "the second card is a spade." $P(A) = 1/13$. To compute $P(B|A)$, we note that if A has occurred then only 12 of the remaining 51 cards are spades, so $P(B|A) = 12/51$ and

$$P(A \cap B) = P(A)P(B|A) = \frac{13}{52} \cdot \frac{12}{51}$$

Note that in this example we computed $P(B|A)$ by thinking about the situation that exists after A has occurred, rather than using the definition $P(B|A) = P(A \cap B)/P(A)$. Indeed, it is more common to use $P(A)$ and $P(B|A)$ to compute $P(A \cap B)$ than to use $P(A)$ and $P(A \cap B)$ to compute $P(B|A)$.

Example 3.5

The Cornell hockey team is playing in a four-team tournament. In the first round they have any easy opponent that they will beat 80% of the time but if they win that game they will play against a tougher team where their probability of success is 0.4. What is the probability that they will win the tournament?

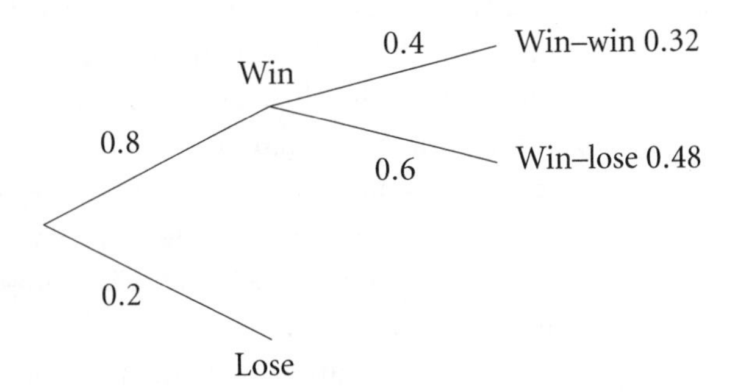

If A and B are the events of victory in the first and second games then $P(A) = 0.8$ and $P(B|A) = 0.4$, so the probability that they will win the tournament is

$$P(A \cap B) = P(A)P(B|A) = 0.8(0.4) = 0.32$$

The reasoning in the last two examples extends easily to three events:

$$P(A_1 \cap A_2 \cap A_3) = P(A_1)P(A_2|A_1)P(A_3|A_1 \cap A_2)$$

since the right-hand side is equal to

$$P(A_1) \cdot \frac{P(A_1 \cap A_2)}{P(A_1)} \cdot \frac{P(A_1 \cap A_2 \cap A_3)}{P(A_1 \cap A_2)}$$

Example 3.6

In the town of Mythica 90% of students graduate high school, 60% of high school graduates complete college, and 20% of college graduates get graduate or professional degrees. What fraction of students get advanced degrees?

Answer $= (0.9)(0.6)(0.2) = 0.108$.

The formula for three events generalizes to any number of events.

Example 3.7

What is the probability of a flush, that is, all cards of the same suit, when we draw 5 cards out of a deck of 52?

$$1 \cdot \frac{12}{51} \cdot \frac{11}{50} \cdot \frac{10}{49} \cdot \frac{9}{48}$$

The first time we can draw anything. On the second draw we must pick 1 of the other 12 cards in that suit among the 51 that remain. If we succeed on the second draw then there are 11 good cards out of 50, etc.

Conditional probabilities are the sources of many "paradoxes" in probability. One of these attracted worldwide attention in 1990 when Marilyn vos Savant discussed it in her weekly column in the Sunday's *Parade* magazine.

Example 3.8

The Monty Hall problem. The problem is named for the host of the television show *Let's Make A Deal,* in which contestants were often placed in situations such as the following: Three curtains are numbered 1, 2, and 3. Behind one curtain is a car; behind the other two curtains are donkeys. You pick a curtain, say #1. To build some suspense the host opens up one of the two remaining curtains, say #3, to reveal a donkey. What is the probability you will win given that there is a donkey behind #3? Should you switch curtains and pick #2 if you are given the chance?

Many people argue that "the two unopened curtains are the same, so they each will contain the car with probability 1/2, and hence there is no point in switching." As we will now show, this naive reasoning is incorrect. To compute the answer, we will suppose that the host always chooses to show you a donkey and picks at random if there are two unchosen curtains with donkeys. Assuming you pick curtain #1, there are three possibilities:

	#1	#2	#3	Host's action
Case 1	Donkey	Donkey	Car	Opens #2
Case 2	Donkey	Car	Donkey	Opens #3
Case 3	Car	Donkey	Donkey	Opens #2 or #3

Now $P(\text{case 2, open door \#3}) = 1/3$ and

$$P(\text{case 3, open door \#3}) = P(\text{case 3})\,P(\text{open door \#3}|\text{case 3}) = \frac{1}{3} \cdot \frac{1}{2} = \frac{1}{6}$$

Adding the two ways door #3 can be opened gives $P(\text{open door \#3}) = 1/2$ and it follows that

$$P(\text{case 3}|\text{open door \#3}) = \frac{P(\text{case 3, open door \#3})}{P(\text{open door \#3})} = \frac{1/6}{1/2} = \frac{1}{3}$$

Although it took a number of steps to compute this answer, it is "obvious." When we picked one of the three doors initially we had probability 1/3 of picking the car, and since the host can always open a door with a donkey, the new information does not change our chance of winning.

The paradox actually predates the game show in the following form. Three prisoners, Al, Bob, and Charlie, are in a cell. At dawn two will be set free and one will be hanged, but they do not know who will be chosen. The guard offers to tell Al the name of one of the other two prisoners who will go free but Al stops him, screaming, "No, don't! That would increase my chances of being hanged to 1/2."

Example 3.9

Cognitive dissonance. An economist, M. Keith Chen, has recently uncovered a version of the Monty Hall problem in the theory of cognitive dissonance. For a half century, experimenters have been using the so-called free choice paradigm to test our tendency to rationalize decisions. In an experiment typical of the genre, Yale psychologists measured monkeys' preferences by observing how quickly each monkey sought out different colors of M&Ms.

In the first step, the researchers gave the monkey a choice between, say, red and blue. If the monkey chose red then it was given a choice between blue and green. Nearly two-thirds of the time it rejected blue in favor of green, which seemed to jibe with the theory of choice rationalization: once we reject something, we tell ourselves we never liked it anyway.

Putting aside this interpretation it is natural to ask: What would happen if monkeys were acting at random? If so, the six orderings RGB, RBG, GRB, GBR, BGR, and BRG would have equal probability. In the first three cases red is preferred to blue, but in 2/3s of those cases green is preferred to blue. Just as in the Monty Hall problem, we think that the probability of preferring blue to green is 1/2 due to symmetry, but the probability is 1/3. This time however conditioning on red being preferred to green reduced the original probability of 1/2 to 1/3, whereas in the Monty Hall problem the probability was initially 1/3 and did not change.

3.2 Two-stage experiments

We begin with several examples and then describe the collection of problems we treat in this section.

Example 3.10 An urn contains 5 red and 10 black balls. We draw 2 balls from the urn without replacement. What is the probability that the second ball drawn is red?

This is easy to see if we draw a picture. The first split in the tree is based on the outcome of the first draw and the second on the outcome of the final. The outcome of the first draw dictates the probabilities for the second one. We multiply the probabilities on the edges to get probabilities of the four endpoints and then sum the ones that correspond to red to get the answer: $4/42 + 10/42 = 1/3$.

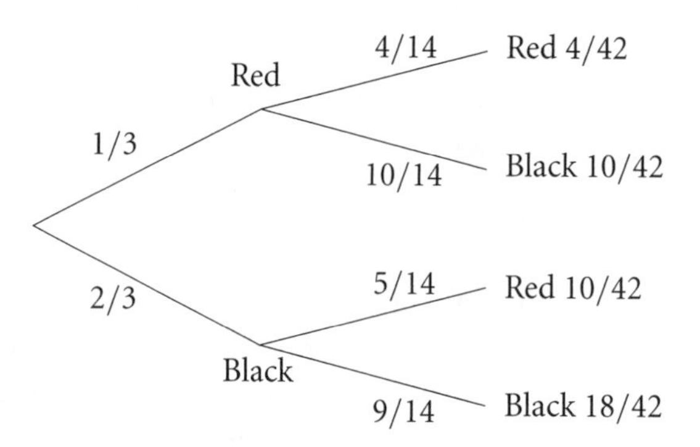

To do this with formulas, let R_i be the event of a red ball on the ith draw and B_1 the event of a black ball on the first draw. Breaking things down according to the outcome of the first test, then using the multiplication rule, we have

$$\begin{aligned} P(R_2) &= P(R_2 \cap R_1) + P(R_2 \cap B_1) \\ &= P(R_2|R_1)P(R_1) + P(R_2|B_1)P(B_1) \\ &= (1/3)(4/14) + (2/3)(5/14) = 14/42 = 1/3 \end{aligned}$$

From this we see that $P(R_2|R_1) < P(R_1) < P(R_2|B_1)$ but the two probabilities average to give $P(R_1)$. This calculation makes the result look like a miracle but it is not. If we number the 15 balls in the urn then by symmetry each of them is equally likely to be the second ball chosen. Thus the probability of a red on the second, eighth, or fifteenth draw is always the same.

Example 3.11

Based on past experience, 70% of students in a certain course pass the midterm exam. The final exam is passed by 80% of those who passed the midterm, but only by 40% of those who fail the midterm. What fraction of students pass the final?

Drawing a tree as before with the first split based on the outcome of the midterm and the second on the outcome of the final, we get the answer: $0.56 + 0.12 = 0.68$.

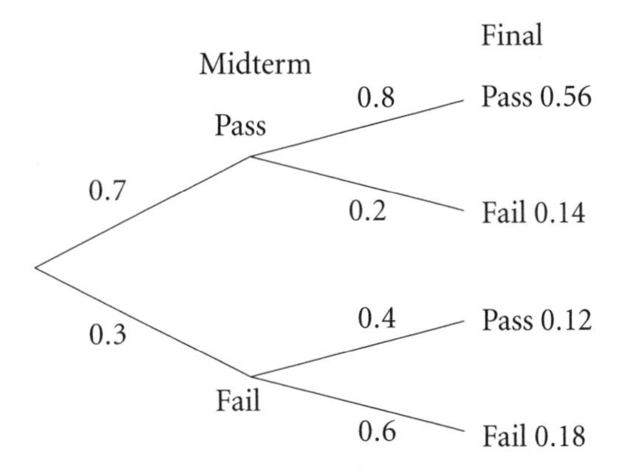

To do this with formulas, let A be the event that the student passes the final and B the event that the student passes the midterm. Breaking things down according to the outcome of the first test, then using the multiplication rule, we have

$$\begin{aligned} P(A) &= P(A \cap B) + P(A \cap B^c) \\ &= P(A|B)P(B) + P(A|B^c)P(B^c) \\ &= (0.8)(0.7) + (0.4)(0.3) = 0.68 \end{aligned}$$

Example 3.12

Al flips 3 coins and Betty flips 2. Al wins if the number of heads he gets is more than the number Betty gets. What is the probability that Al will win?

Let W be the event that Al wins. We will break things down according to the number of heads Betty gets. Let B_i be the event that Betty gets i heads and A_j the event that Al gets j heads. By considering the 4 outcomes of flipping 2 coins it is easy to see that

$$P(B_0) = 1/4 \qquad P(B_1) = 1/2 \qquad P(B_2) = 1/4$$

while considering the 8 outcomes for 3 coins leads to

$$P(W|B_0) = P(A_1 \cup A_2 \cup A_3) = 7/8$$
$$P(W|B_1) = P(A_2 \cup A_3) = 4/8$$
$$P(W|B_2) = P(A_3) = 1/8$$

This gives us the raw material for drawing our picture:

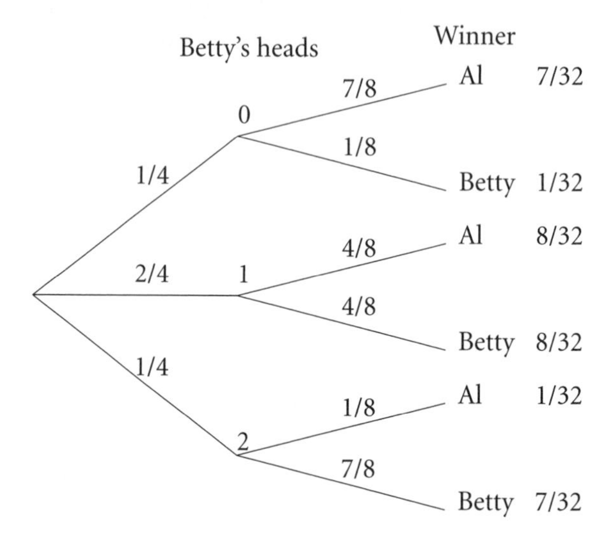

Adding up the ways Al can win, we get $7/32 + 8/32 + 1/32 = 1/2$. To check this draw a line through the middle of the picture and note the symmetry between top and bottom.

To do this with formulas, note that $W \cap B_i$, $i = 0, 1, 2$, are disjoint and their union is W, so

$$P(W) = \sum_{i=0}^{2} P(W \cap B_i) = \sum_{i=0}^{2} P(W|B_i)P(B_i)$$

since $P(W \cap B_i) = P(A|B_i)P(B_i)$ by the multiplication rule (3.3). Plugging in the values we computed,

$$P(W) = \frac{1}{4} \cdot \frac{7}{8} + \frac{2}{4} \cdot \frac{4}{8} + \frac{1}{4} \cdot \frac{1}{8} = \frac{7+8+1}{32} = \frac{1}{2}$$

The previous analysis makes it look miraculous that we have a fair game. However, it is true in general.

Example 3.13

Al flips $n+1$ coins and Betty flips n. Al wins if the number of heads he gets is more than the number Betty gets. What is the probability that Al will win?

Consider the situation after Al has flipped n coins and Betty has flipped n. Using X and Y to denote the number of heads for Al and Betty at that time, there are the three possibilities: $X > Y, X = Y, X < Y$. In the first case, Al has already won. In the third, he cannot win. In the second, he wins with probability 1/2. Using symmetry if $P(X > Y) = P(X < Y) = p$ then $P(X = Y) = 1 - 2p$, so the probability that Al wins is $p + (1 - 2p)/2 = 1/2$.

Abstracting the structure of the last problem, let $B_1, \ldots, B_k$ be a **partition**, that is, a collection of disjoint events whose union is Ω.

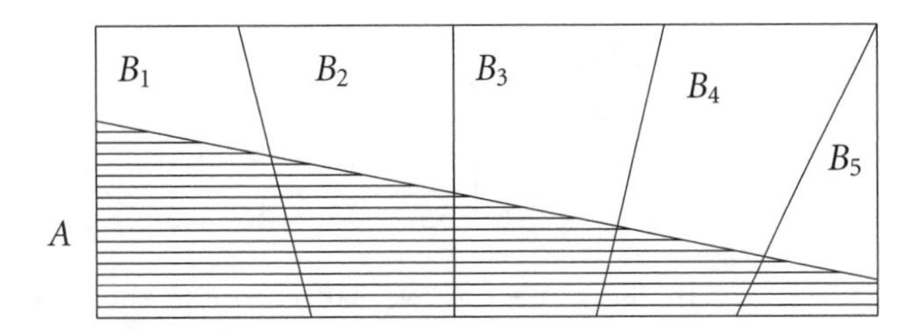

Using the fact that the sets $A \cap B_i$ are disjoint, and the multiplication rule, we have

$$P(A) = \sum_{i=1}^{k} P(A \cap B_i) = \sum_{i=1}^{k} P(A|B_i)P(B_i) \tag{3.4}$$

a formula that is sometimes called the **law of total probability**.

The name of this section comes from the fact that we think of our experiment as occurring in two stages. The first stage determines which of the B's occur, and when B_i occur in the first stage, A occurs with probability $P(A|B_i)$ in the second. As the next example shows, the two stages are sometimes clearly visible in the problem itself.

Example 3.14 Roll a die and then flip that number of coins. What is the probability of $A =$ "we get exactly 3 heads"?

Let $B_i =$ "the die shows i." $P(B_i) = 1/6$ for $i = 1, 2, \ldots, 6$ and

$$P(A|B_1) = 0 \qquad P(A|B_2) = 0 \qquad P(A|B_3) = 2^{-3}$$
$$P(A|B_4) = C_{4,3}\,2^{-4} \qquad P(A|B_5) = C_{5,3}\,2^{-5} \qquad P(A|B_6) = C_{6,3}\,2^{-6}$$

So plugging into (3.4),

$$\begin{aligned} P(A) &= \frac{1}{6}\left\{\frac{1}{8} + \frac{4}{16} + \frac{10}{32} + \frac{20}{64}\right\} \\ &= \frac{1}{6}\left\{\frac{8 + 16 + 20 + 20}{64}\right\} = \frac{1}{6} \end{aligned}$$

Example 3.15

Suppose we roll three dice. What is the probability that the sum is 9?

Let $A =$ "the sum is 9," $B_i =$ "the first die shows i," and $C_j =$ "the sum of the second and third dice is j." Now $P(A|B_i) = P(C_{9-i})$ and we know the probabilities for the sum of two dice:

j	2	3	4	5	6	7	8	9	10	11	12
$P(C_j)$	$\frac{1}{36}$	$\frac{2}{36}$	$\frac{3}{36}$	$\frac{4}{36}$	$\frac{5}{36}$	$\frac{6}{36}$	$\frac{5}{36}$	$\frac{4}{36}$	$\frac{3}{36}$	$\frac{2}{36}$	$\frac{1}{36}$

Using (3.4), now we have

$$P(A) = \sum_{i=1}^{6} P(B_i)P(A|B_i) = \frac{1}{6}\left(P(C_8) + P(C_7) + \cdots + P(C_3)\right)$$

$$= \frac{1}{6}\left(\frac{5}{36} + \frac{6}{36} + \frac{5}{36} + \frac{4}{36} + \frac{3}{36} + \frac{2}{36}\right) = \frac{25}{216}$$

In the same way we can compute the probability of $A_k =$ "The sum of three dice is k." To check the symmetry in the table, note that if the numbers on top are $i_1 + i_2 + i_3 = k$, then the sum of the numbers on the bottom are $(7 - i_1) + (7 - i_2) + (7 - i_3) = 21 - k$.

k	3, 18	4, 17	5, 16	6, 15	7, 14	8, 13	9, 12	10, 11
$P(A_k)$	$\frac{1}{216}$	$\frac{3}{216}$	$\frac{6}{216}$	$\frac{10}{216}$	$\frac{15}{216}$	$\frac{21}{216}$	$\frac{25}{216}$	$\frac{27}{216}$

The graph in the figure shows the shape of the distribution. Note that the triangular shape of the sum of two dice has become a little more rounded.

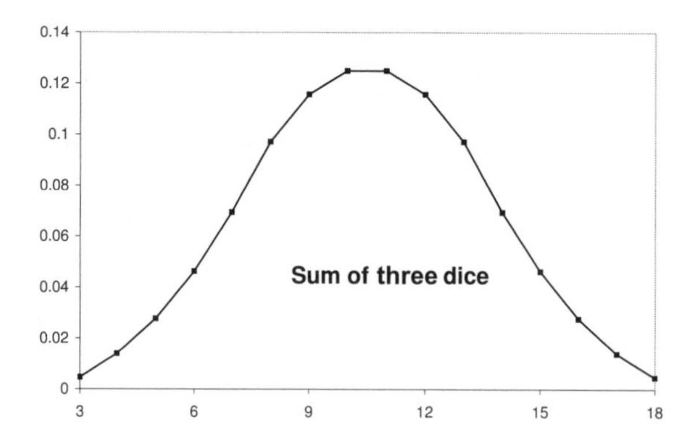

Example 3.16

Craps. In this game, if the sum of the dice is 2, 3, or 12 on his first roll, the player loses; if the sum is 7 or 11, he wins; if the sum is 4, 5, 6, 8, 9, or 10, this number becomes his "point" and he wins if he "makes his point" that is, his number comes up again before he throws a 7. What is the probability that the player wins?

The first step in analyzing craps is to compute the probability that the player makes his point. Suppose his point is 5 and let E_k be the event that the sum is k. There are 4 outcomes in E_5 ((1, 4), (2, 3), (3, 2), (4, 1)), 6 in E_7, and hence 26 not in $E_5 \cup E_7$. Letting $\times$ stand for "the sum is not 5 or 7," we see that

$$P(5) = \frac{4}{36} \qquad P(\times\, 5) = \frac{26}{36} \cdot \frac{4}{36} \qquad P(\times \times 5) = \left(\frac{26}{36}\right)^2 \frac{4}{36}$$

From the first three terms it is easy to see that for $k \geq 0$,

$$P(\times \text{ on } k \text{ rolls then } 5) = \left(\frac{26}{36}\right)^k \frac{4}{36}$$

Summing over the possibilities, which represent disjoint ways of rolling 5 before 7, we have

$$P(5 \text{ before } 7) = \sum_{k=0}^{\infty} \left(\frac{26}{36}\right)^k \frac{4}{36} = \frac{4}{36} \cdot \frac{1}{1 - \frac{26}{36}}$$

since

$$\sum_{k=0}^{\infty} x^k = \frac{1}{1-x} \tag{3.5}$$

Simplifying, we have $P(5 \text{ before } 7) = (4/36)/(10/36) = 4/10$. Such a simple answer should have a simple explanation, and it does. Consider an urn with four balls marked 5, six marked 7, and twenty-six marked with x. Drawing with replacement until we draw either a 5 or 7 is the same as drawing once from an urn with 10 balls with four balls marked 5 and six marked 7.

5 5 5 5 x x x x x x x x
7 7 7 7 7 7 x x x x x x
x x x x x x x x x x x x

Another way of saying this is that if we ignore the outcomes that result in a sum other than 5 or 7, we reduce the sample space from Ω to $E = E_5 \cup E_7$ and the distribution of the first outcome that lands in E follows the conditional probability $P(\cdot|E)$. Since $E_5 \cap E = E_5$, we have

$$P(E_5|E) = \frac{P(E_5)}{P(E)} = \frac{4/36}{10/36} = \frac{4}{10}$$

The last argument generalizes easily to give the probabilities of making any point:

k	4	5	6	8	9	10
$\|E_k\|$	3	4	5	5	4	3
P(k before 7)	3/9	4/10	5/11	5/11	4/10	3/9

To compute the probability of A = "he wins," we let B_k = "the first roll is k" and observe that (3.4) implies

$$P(A) = \sum_{k=2}^{12} P(A \cap B_k) = \sum_{k=2}^{12} P(B_k)P(A|B_k)$$

When $k = 2$, 3, or 12 comes up on the first roll we lose, so

$$P(A|B_k) = 0 \quad \text{and} \quad P(A \cap B_k) = 0$$

When $k = 7$ or 11 comes up on the first roll we win, so

$$P(A|B_k) = 1 \quad \text{and} \quad P(A \cap B_k) = P(B_k)$$

When the first roll is $k = 4, 5, 6, 8, 9,$ or 10, $P(A|B_k) = P(k \text{ before } 7)$ and $P(A \cap B_k)$ is

$$\frac{3}{36} \cdot \frac{3}{9} \quad k = 4, 10 \qquad \frac{4}{36} \cdot \frac{4}{10} \quad k = 5, 9 \qquad \frac{5}{36} \cdot \frac{5}{11} \quad k = 6, 8$$

Adding up the terms in the sum in the order in which they were computed,

$$\begin{aligned} P(A) &= \frac{6}{36} + \frac{2}{36} + 2\left(\frac{1}{36} + \frac{4 \cdot 2}{36 \cdot 5} + \frac{5 \cdot 5}{36 \cdot 11}\right) \\ &= \frac{4}{18} + 2\left(\frac{55 + 88 + 125}{36 \cdot 11 \cdot 5}\right) = \frac{220 + 268}{18 \cdot 11 \cdot 5} = \frac{488}{990} = 0.4929 \end{aligned} \tag{3.6}$$

which is not very much less than $1/2 = 495/990$.

Example 3.17 Al and Bob take turns throwing one dart to try to hit a bull's-eye. Al hits with probability 1/4, while Bob hits with probability 1/3. If Al goes first what is the probability that, he will hit the first bull's-eye?

Let p be the answer. By considering one cycle of the game, we see

$$p = 1/4 + (3/4)(1/3)(0) + (3/4)(2/3)p$$

In words, Al wins if he hits the bull's-eye on the first try. If he misses and Al hits then he loses. If they both miss then it is Al's turn and the game starts over, so Al's probability of success is p. Solving the equation we have $p/2 = 1/4$ or $p = 1/2$.

Back to craps. This reasoning in the last example can be used to compute the probability q that a player rolls a 5 before 7. By considering the outcome of the first roll $q = 4/36 + (6/36)0 + (26/36)q$ and solving we have $q = 4/10$.

Example 3.18

NCAA Basketball Tournament. Since 1985 the tournament has had 64 teams, four regions with 16 seeded teams. This is a knockout tournament; that is, after each game the loser is eliminated. The table given next presents data for 1985–2004, 20 seasons. Since there are four regions, this means that each seeding has had a total of 80 trials. The table describes relative success of the various seeds in advancing in the tournament to the rounds of 32, sweet 16, elite 8, the final 4, the 2 teams in the championship game, and to win the tournament. The numbers are decreasing across each row. For readability once a number becomes 0, the remaining entries are left blank.

For reasons that will become clear as you read the table we have listed the seeds in the order dictated by how the games are played. That is, in the first round the sum of the seeds of the two teams is always 17, and the number of times the teams advance will add up to 80. In the round of 16 statistics if we divide the 16 numbers into four groups of four, each will add up to 80, etc.

Seed	32	16	8	4	2	Winner
1	80	68	56	34	17	11
16	0					
8	37	9	6	3	1	1
9	43	3	1	0		
4	64	36	12	7	2	1
13	16	3	0			
5	54	28	4	3	2	0
12	26	13	1	0		
3	67	38	18	11	7	2
14	13	2	0			
6	56	30	11	3	2	1
11	24	10	3	1	0	
2	76	51	37	18	9	4
15	4	0				
7	48	13	5	0		
10	32	16	6	0		
Total	640	320	160	80	40	20

From this table we can compute the probabilities for the first four seeds to win a game in each round, given that it reached that round.

	64	32	16	8	4	2
1	1.0	0.85	0.823	0.607	0.5	0.647
2	0.95	0.671	0.725	0.486	0.5	0.444
3	0.838	0.567	0.474	0.611	0.636	0.285
4	0.8	0.563	0.333	0.583	0.286	0.5

Here, $68/80 = 0.85$, $56/68 = 0.823$, etc. As we should expect, the conditional probabilities generally decrease from left to right and from top to bottom. We leave it to the reader to ponder the meaning of the exceptions, some of which may be due only to the small sample sizes.

3.3 Bayes' formula

The title of the section is a little misleading since we will regard Bayes' formula as a method for computing conditional probabilities and will only reluctantly give the formula after we have done several examples to illustrate the method.

Example 3.19

Exit polls. In the California gubernatorial election in 1982, several TV stations predicted, on the basis of questioning people when they exited the polling place, that Tom Bradley, then mayor of Los Angeles, would win the election. When the votes were counted, however, he lost to George Deukmejian by a considerable margin. What caused the exit polls to be wrong?

To give our explanation we need some notation and some numbers. Suppose we choose a person at random; let $B =$ "the person votes for Bradley" and suppose that $P(B) = 0.45$. There were only two candidates, so this makes the probability of voting for Deukmejian $P(B^c) = 0.55$. Let $A =$ "the voter stops and answers a question about how they voted," and suppose that $P(A|B) = 0.4$ and $P(A|B^c) = 0.3$. That is, 40% of Bradley voters will respond compared with 30% of the Deukmejian voters. We are interested in computing $P(B|A) =$ the fraction of voters in our sample that voted for Bradley. By the definition of conditional probability (1.6),

$$P(B|A) = \frac{P(B \cap A)}{P(A)} = \frac{P(B \cap A)}{P(B \cap A) + P(B^c \cap A)}$$

To evaluate the two probabilities, we use the multiplication rule (3.3):

$$P(B \cap A) = P(B)P(A|B) = 0.45 \cdot 0.4 = 0.18$$

$$P(B^c \cap A) = P(B^c)P(A|B^c) = 0.55 \cdot 0.3 = 0.165$$

From this it follows that

$$P(B|A) = \frac{0.18}{0.18 + 0.165} = 0.5217$$

and from our sample it looks as if Bradley will win. The problem with the exit poll is that the difference in the response rates makes our sample not representative of the population as a whole.

Turning to the mechanics of the computation, note that 18% of the voters are for Bradley and respond, while 16.5% are for Deukmejian and respond, so the fraction of Bradley voters in our sample is $18/(18 + 16.5)$. In words, there are two ways an outcome can be in A – it can be in B or in B^c – and the conditional probability is the fraction of the total that comes from the first way.

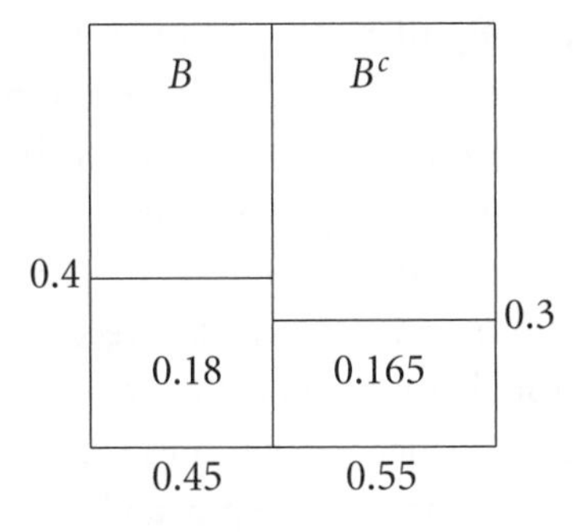

Example 3.20

Mammogram posterior probabilities. Approximately 1% of women aged 40–50 years have breast cancer. A woman with breast cancer has a 90% chance of a positive test from a mammogram, while a woman without has a 10% chance of a false-positive result. What is the probability that a woman has breast cancer given that she just had a positive test?

Let B = "the woman has breast cancer" and A = "a positive test." We want to calculate $P(B|A)$. Computing as in the previous example,

$$P(B|A) = \frac{P(B \cap A)}{P(A)} = \frac{P(B \cap A)}{P(B \cap A) + P(B^c \cap A)}$$

To evaluate the two probabilities, we use the multiplication rule (3.3):

$$P(B \cap A) = P(B)P(A|B) = 0.01 \cdot 0.9 = 0.009$$

$$P(B^c \cap A) = P(B^c)P(A|B^c) = 0.99 \cdot 0.1 = 0.099$$

From this it follows that

$$P(B|A) = \frac{0.009}{0.009 + 0.099} = \frac{9}{108}$$

or a little less than 9%. This situation comes about because it is much easier to have a positive result from a false-positive for a healthy woman, which has probability 0.099, than from a woman with breast cancer having a positive test, which has probability 0.009.

This answer is somewhat surprising. Indeed, when 95 physicians were asked the question "What is the probability a woman has breast cancer given that she just had a positive test?", their average answer was 75%. The two statisticians who carried out this survey indicated that physicians were better able to see the answer when the data were presented in frequency format. Ten out of 1,000 women have breast cancer. Of these 9 will have a positive mammogram. However, of the remaining 990 women without breast cancer, 99 will have a positive test, and again we arrive at the answer $9/(9+99)$.

Example 3.21

Hemophilia. Ellen has a brother with hemophilia, but has two parents who do not have the disease. Since hemophilia is caused by a recessive allele h on the X chromosome, we can infer that her mother is a carrier (that is, the mother has the hemophilia allele h on one of her X chromosomes and the healthy allele H on the other), while her father has the healthy allele on his one X chromosome. Since Ellen received one X chromosome from her father and one from her mother, there is a 50% chance that she is a carrier, and if so, there is a 50% chance that her sons will have the disease. If she has two sons without the disease, what is the probability that she is a carrier?

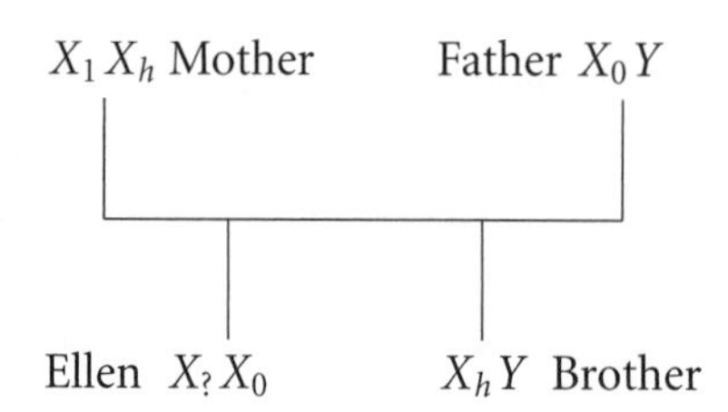

Let B be the event that she is a carrier and A the event that she has two healthy sons. Computing as in the two previous examples,

$$P(B|A) = \frac{P(B\cap A)}{P(A)} = \frac{P(B\cap A)}{P(B\cap A)+P(B^c\cap A)}$$

To evaluate the two probabilities we use the multiplication rule (3.3). Since the probability of having two healthy sons when she is a carrier is 1/4, and is 1 when she is not.

$$P(B\cap A) = P(B)P(A|B) = \frac{1}{2}\cdot\frac{1}{4} = \frac{1}{8}$$

$$P(B^c\cap A) = P(B^c)P(A|B^c) = \frac{1}{2}\cdot 1 = \frac{1}{2}$$

From this it follows that

$$P(B|A) = \frac{1/8}{1/8 + 1/2} = \frac{1}{5}$$

Example 3.22 Three factories make 20, 30, and 50% of the computer chips for a company. The probability of a defective chip is 0.04, 0.03, and 0.02 for the three factories. We have a defective chip. What is the probability that it came from Factory 1?

Let B_i be the event that the chip came from factory i and A the event that the chip is defective. We want to compute $P(B_3|A)$. Adapting the computation from the two previous examples to the fact that there are now three B_i,

$$P(B|A) = \frac{P(B \cap A)}{P(A)} = \frac{P(B \cap A)}{\sum_{i=1}^{3} P(B_i \cap A)}$$

To evaluate the three probabilities, we use the multiplication rule (3.3):

$$P(B_1 \cap A) = P(B_1)P(A|B_1) = 0.2 \cdot (0.04) = 0.008$$

$$P(B_2 \cap A) = P(B_2)P(A|B_2) = 0.3 \cdot (0.03) = 0.009$$

$$P(B_3 \cap A) = P(B_3)P(A|B_3) = 0.5 \cdot (0.02) = 0.010$$

From this it follows that

$$P(B_1|A) = \frac{P(B_1 \cap A)}{P(A)} = \frac{0.008}{0.008 + 0.009 + 0.010} = \frac{8}{27}$$

The calculation can be summarized by the following picture. The conditional probability $P(B_3|A)$ is the fraction of the event A that lies in B_3.

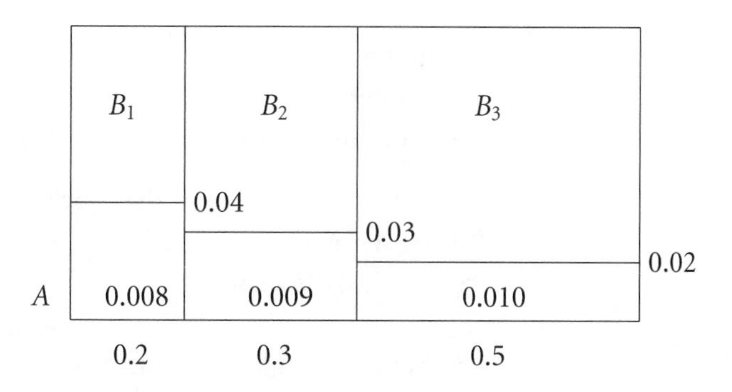

We are now ready to generalize from our examples and state **Bayes' formula.** In each case, we have a **partition** of the probability space $B_1, \ldots, B_n$, that is, a sequence of disjoint sets with $\cup_{i=1}^{n} B_i = \Omega$. (In the first three examples, $B_1 = B$ and $B_2 = B^c$.) We are given $P(B_i)$ and $P(A|B_i)$ for $1 \le i \le n$ and we want to compute $P(B_1|A)$. Reasoning as in the previous examples,

$$P(B_1|A) = \frac{P(B_1 \cap A)}{P(A)} = \frac{P(B_1 \cap A)}{\sum_i P(B_i \cap A)}$$

To evaluate the probabilities, we observe that

$$P(B_i \cap A) = P(B_i)P(A|B_i)$$

From this, it follows that

$$P(B_1|A) = \frac{P(B_1 \cap A)}{P(A)} = \frac{P(B_1)P(A|B_1)}{\sum_{i=1}^{n} P(B_i)P(A|B_i)} \tag{3.7}$$

This is Bayes' formula. Even though we have numbered it, we advise you not to memorize it. It is much better to remember the procedures we followed to compute the conditional probability.

Our last two examples come from law.

Example 3.23

Paternity probabilities. Before there were sophisticated tests based on DNA samples, the testing of blood type and other hereditary factors was used in paternity cases to infer, using Bayes' formula, the probability that a particular man is the father. For a concrete example, suppose that the baby's blood type is B, the mother's is A, and that of the suspected father's, whom for convenience we will call Bob, is B. Given this information what is the probability that Bob is the father?

To explain how this could happen, we note that the genes that control blood type can be O, A, or B, with A and B dominant over O but neither A nor B dominating the other, so we get the following correspondence between genotypes (the genes on the two chromosomes) and phenotypes (observed blood type):

Genotype	OO	AO	AA	BO	BB	AB
Phenotype	O	A	A	B	B	AB
Proportion	0.479	0.310	0.050	0.116	0.007	0.038

From this table, we see that if the baby's blood type is B then it must be the case that the mother's genotype is AO, she contributed an O gene, and the father contributed a B gene.

Let E (for "evidence") be the event that the baby's blood type is B and F the event that Bob is indeed the father. We cannot observe Bob's genotype, but using the proportions of the various genotypes from the table, we can compute that

$$P(\text{genotype is } BO|\text{ phenotype is } B) = 0.116/0.123$$

$$P(E|F) = \frac{(0.116)0.5 + 0.007}{0.123} = \frac{0.065}{0.123} = 0.528$$

There is not too much to argue about in the last computation. When we compute $P(E|F^c)$, we make the first of two questionable assumptions: If Bob is not the father, then the real father is someone chosen at random from the population, so

$$P(E|F^c) = (0.116)0.5 + 0.007 + (0.038)0.5 = 0.084$$

To evaluate $P(F|E)$, we thus need to evaluate the **prior probability** $P(F)$ that Bob is the father. It would be natural to make $P(F)$ equal to the fraction of times that the mother had intercourse with Bob near the time of conception. However, it is not unusual for the mother to claim that this number is 1 and the alleged father to claim that it is 0, so the common practice in these computations is to set $P(F) = 1/2$ (our second questionable assumption). If we do this, then $P(F) = P(F^c) = 1/2$, so

$$\begin{aligned} P(F|E) &= \frac{P(E|F)P(F)}{P(E|F)P(F) + P(E|F^c)P(F^c)} \\ &= \frac{0.065/0.123}{0.065/0.123 + 0.084} = \frac{1}{1.158} = 0.8628 \end{aligned}$$

Example 3.24

O. J. Simpson trial. DNA testing has considerably more power than blood tests. RFLPs (restriction fragment length polymorphisms) are typed by digesting DNA with "restriction" enzymes and then determining the lengths of the fragments. These lengths are highly variable (polymorphic) in humans, so the use of eight or nine such markers results in incredibly small probabilities. For example, Robin Cotton of Cellmark Diagnostics testified that blood found on a sock near Simpson's bed had the genetic type of Nicole Brown Simpson and the chances of another person having the exact same RFLP alleles were 1 in 9.7 billion.

For this problem we will concentrate on a less dramatic example that involves blood found at the murder scene (item 49). Three blood factors were recorded that matched Simpson's blood types. The next table that comes from p. 10 of Vol. 7, No. 4, of *Chance* gives the frequencies estimated from the overall population.

System	Item #49	Frequency
ABO	A	0.347
EsD	1	0.79
PGM	2+, 2−	0.016

Here, 2+, 2− indicates that two alleles were present, one inherited from each parent.

Multiplying the probabilities together gives 0.00438 or 1/227, a number that was approximated in the trial and quoted in the press as 1/200. Letting E denote this evidence and G the event that Simpson is guilty, and sticking with the simpler fraction, we see that $P(E|G^c) = 1/200$. It is an error known as the "prosecutor's fallacy" to think of this as $P(G^c|E)$; that is, the probability that Simpson is innocent given this evidence is 1/200. A second error known as the "defendant's fallacy" is to note that 1/200 of the population of Los Angeles is 40,000, so the probability that it is O. J. Simpson's blood is 1/40,000.

Both these fallacies are based on assuming that unknown probabilities are uniform on the set of possibilities. The correct way to compute $P(G|E)$ is

$$P(G|E) = \frac{P(E|G)P(G)}{P(E|G)P(G) + P(E|G^c)P(G^c)}$$

but of course this requires giving a value to $P(G)$. It is perhaps for this reason that Bruce Weir (*Nature Genetics*, Vol. 11, pp. 365–368) argues for the use of the likelihood ratio $P(E|G)/P(E|G^c) = 200$, that is, the evidence is 200 times more likely if O. J. Simpson is guilty than if the murderer is a randomly chosen person.

3.4 Discrete joint distributions

In many situations we need to know the relationship between several random variables $X_1, \ldots, X_n$. Here, we confine our attention to the case $n = 2$. Once this case is understood the extension to $n > 2$ is straightforward. For one random variable, the distribution is a list of the probabilities of all the possible values. The joint distribution of a pair of random variables is a table of numbers that gives the probabilities for all the possible values of the pair.

Example 3.25 Roll 2 four-sided dice with the numbers 1, 2, 3, and 4 on their sides. Let X be the maximum of the two numbers that appear and Y the sum.

By considering the 16 possible outcomes we find the following joint distribution of X and Y.

X	$Y = 2$	3	4	5	6	7	8
1	1/16	0	0	0	0	0	0
2	0	2/16	1/16	0	0	0	0
3	0	0	2/16	2/16	1/16	0	0
4	0	0	0	2/16	2/16	2/16	1/16

Example 3.26 Suppose we draw 2 balls out of an urn with 6 red, 5 blue, and 4 green balls. Let X be the number of red balls we get and Y the number of blue balls.

There are $C_{15,2} = 105$ outcomes. The number of outcomes with i red, j blue, and $2 - (i + j)$ green balls is $C_{6,i}C_{5,j}C_{4,2-(i+j)}$. Using this formula, we have

X	$Y = 0$	1	2
0	$\frac{1\cdot1\cdot6}{105} = 6/105$	$\frac{1\cdot5\cdot4}{105} = 20/105$	$\frac{1\cdot10\cdot1}{105} = 10/105$
1	$\frac{6\cdot1\cdot4}{105} = 24/105$	$\frac{6\cdot5\cdot1}{105} = 30/105$	0
2	$\frac{15\cdot1\cdot1}{105} = 15/105$	0	0

Example 3.27 Consider the following hypothetical joint distribution of X, a person's grade on the AP calculus exam (a number between 1 and 5), and their grade Y in their high school calculus course, which we assume was $A = 4$, $B = 3$, or $C = 2$.

X	$Y = 4$	3	2
5	0.1	0.05	0
4	0.15	0.15	0
3	0.1	0.15	0.10
2	0	0.05	0.10
1	0	0	0.05

3.4.1 Marginal distributions

The next question to be addressed is, "Given the joint distribution of (X, Y), how do we recover the distributions of X and Y?" The answer is that the **marginal distributions** of X and Y are given by

$$\begin{aligned} P(X = x) &= \textstyle\sum_y P(X = x, Y = y) \\ P(Y = y) &= \textstyle\sum_x P(X = x, Y = y) \end{aligned} \tag{3.8}$$

To explain the first formula in words, if $X = x$, then Y will take on some value y; so to find $P(X = x)$, we sum the probabilities of the disjoint events $\{X = x, Y = y\}$ over all the values of y. To illustrate these formulas we return to Example 3.25, where we rolled a four-sided die and let X be the larger number and Y the sum. Omitting the probabilities that are 0,

$$\begin{aligned} P(X = 1) &= P(X = 1, Y = 1) = 1/16 \\ P(X = 2) &= P(X = 2, Y = 3) + P(X = 2, Y = 4) = 2/16 + 1/16 = 3/16 \\ P(X = 3) &= P(X = 3, Y = 4) + P(X = 3, Y = 5) + P(X = 3, Y = 6) \\ &= 2/16 + 2/16 + 1/16 = 5/16 \\ P(X = 4) &= \sum_{y=5}^{8} P(X = 4, Y = y) = 2/16 + 2/16 + 2/16 + 1/16 = 7/16 \end{aligned}$$

In words, we add the probabilities in each row to get the marginal distribution of X. Similarly, we add up the probabilities in each column to get the distribution of Y.

X	$Y=2$	3	4	5	6	7	8	
1	1/16	0	0	0	0	0	0	1/16
2	0	2/16	1/16	0	0	0	0	3/16
3	0	0	2/16	2/16	1/16	0	0	5/16
4	0	0	0	2/16	2/16	2/16	1/16	7/16
	1/16	2/16	3/16	4/16	3/16	2/16	1/16	

In the urn example,

X	$Y=0$	1	2	
0	6/105	20/105	10/105	36/105
1	24/105	30/105	0	54/104
2	15/105	0	0	15/105
	45/105	50/105	10/105	

To check the marginal distribution of Y note that when we draw from an urn with 6 red, 5 blue, and 4 green balls, the probabilities for the number of blue balls are

$$0: \frac{C_{10,2}}{105} = \frac{45}{105} \qquad 1: \frac{C_{5,1}C_{10,1}}{105} = \frac{50}{105} \qquad 2: \frac{C_{5,2}}{105} = \frac{10}{105}$$

Finally in the AP calculus example, the marginal distributions are

X	$Y=4$	3	2	
5	0.10	0.05	0	0.15
4	0.15	0.15	0	0.30
3	0.10	0.15	0.10	0.35
2	0	0.05	0.10	0.15
1	0	0	0.05	0.05
	0.35	0.40	0.25	

3.4.2 Independence

Two random variables are **independent** if

$$P(X = x, Y = y) = P(X = x)P(Y = y) \tag{3.9}$$

In words, two random variables are independent if for each x and y, the events $\{X = x\}$ and $\{Y = y\}$ are independent. To use two of our new terms, this occurs if their joint distribution is the product of the two marginal distributions. In the dice example,

$$P(X = 1, Y = 4) = 0 < (1/16) \cdot (3/16) = P(X = 1)P(Y = 4)$$

so the random variables are not independent. In general, independence fails if there is a 0 in the table where the row and column sums are positive. This simple observation takes care of all our examples; so to further explore the concept we need a new one.

X	$Y = 0$	1	
0	0.4	0.3	0.7
1	0.2	0.1	0.3
	0.6	0.4	

There is no zero in the table but it fails the independence test:

$$P(X = 0, Y = 0) = 0.4 \neq 0.42 = (0.7)(0.6) = (X = 0)P(Y = 0)$$

If we want to have independent random variables with these marginal distributions there is only one way to fill in the table.

X	$Y = 0$	1	
0	0.42	0.28	0.7
1	0.18	0.12	0.3
	0.6	0.4	

Example 3.28

This example gives a remarkable property of the Poisson distribution. Let $A_1, \ldots, A_k$ be disjoint events whose union $\cup_{i=1}^k A_i = \Omega$. Suppose we perform the experiment a random number of times N, where N has a Poisson distribution with mean λ, and let X_i be the number of times A_i occurs.

If $n = x_1 + \cdots + x_k$, then recalling the formula for the multinomial distribution (Example 2.9 in Chapter 2),

$$P(X_i = x_i \text{ for } 1 \leq i \leq k) = e^{-\lambda}\frac{\lambda^n}{n!}\frac{n!}{x_1! \cdots x_k!}P(A_1)^{x_1} \cdots P(A_k)^{x_k}$$

$$= e^{-\lambda P(A_1)}\frac{(\lambda P(A_1))^{x_1}}{x_1!} \cdots e^{-\lambda P(A_k)}\frac{(\lambda P(A_k))^{x_k}}{x_k!}$$

since $\sum_{i=1}^k P(A_i) = 1$. In words, $X_1, \ldots, X_k$ are independent Poissons with parameters $\lambda P(A_i)$.

To see why this is surprising, consider the special case $k = 2$; that is, $A_2 = A_1^c$. If we performed our experiment a fixed number of times then N_1 and N_2 would not be independent since $N_2 = n - N_1$. It is remarkable that when we perform our experiment a Poisson number of times, the number of successes tells us nothing about the number of failures. This result is not only surprising but also useful. For a concrete example, suppose that a Poisson number of cars arrive at a fast-food restaurant each hour and let A_i be the event that the car has i passengers. Then the number of cars with i passengers that arrive are independent Poissons.

3.4.3 Conditional distributions

For discrete random variables, the definition of conditional probability implies

$$P(X = x|Y = y) = \frac{P(X = x, Y = y)}{P(Y = y)} = \frac{P(X = x, Y = y)}{\sum_u P(X = u, Y = y)} \tag{3.10}$$

If we fix y and look at $P(X = x|Y = y)$ as a function of x, what we have is the **conditional distribution of X given that $Y = y$.**

Example 3.29 To illustrate this formula we look at our AP calculus example:

X	$Y = 4$	3	2	
5	0.10	0.05	0	0.15
4	0.15	0.15	0	0.30
3	0.10	0.15	0.10	0.35
2	0	0.05	0.10	0.15
1	0	0	0.05	0.05
	0.35	0.40	0.25	

It follows from the definition of conditional probability that

$$P(X = 5|Y = 4) = P(X = 5, Y = 4)/P(Y = 4) = 0.10/0.35 = 2/7$$
$$P(X = 4|Y = 4) = P(X = 4, Y = 4)/P(Y = 4) = 0.15/0.35 = 3/7$$
$$P(X = 3|Y = 4) = P(X = 3, Y = 4)/P(Y = 4) = 0.10/0.35 = 2/7$$

In words, 2/7's of the students who get A's in the course get a 5 on the exam, 3/7's get a 4, and 2/7's get a 3. Operationally, we divide the entries in the second column by their sum to turn them into a probability distribution. We leave it to the reader to check

x	5	4	3	2	1
$P(X = x\|Y = 3)$	1/8	3/8	3/8	1/8	0
$P(X = x\|Y = 2)$	0	0	2/5	2/5	1/5

The conditional expectation is mean of the conditional distribution:

$$E(X|Y = y) = \sum_x xP(X = x|Y = y)$$

In the previous example,

$$E(X|Y = 4) = (5(0.10) + 4(0.15) + 3(0.10))/0.35 = 4$$
$$E(X|Y = 3) = (5(0.05) + 4(0.15) + 3(0.15) + 2(0.05))/0.40 = 3.5$$
$$E(X|Y = 2) = (3(0.10) + 2(0.10) + 1(0.05))/0.25 = 2.2$$

Example 3.30

Simpson's paradox is the phenomenon that means of subgroups can show much different patterns than the mean of the group as a whole. For a real-life example, consider the average SAT verbal score. It was 504 in 1981 and 21 years later in 2002 it was again 504. However, when we break things down by ethnic groups, we see that all of them increased their scores:

	1981	2002
Non-Hispanic whites	519	527
African-Americans	412	431
Mexican-Americans	438	446
Asian-Americans	474	501

The explanation is simple: minorities made up a much larger portion of the testing population in 2002 than in 1981, and although they have shown significant improvement their averages are lower than non-Hispanic whites, which reduces the overall mean.

3.5 Exercises

Conditional probability

1. A friend flips two coins and tells you that at least one is head. Given this information, what is the probability that the first coin is head?

2. A friend rolls two dice and tells you that there is at least one 6. What is the probability the sum is at least 9?

3. Suppose we roll two dice. What is the probability that the sum is 7 given that neither die showed a 6?

4. Suppose you draw 5 cards out of a deck of 52 and get 2 spades and 3 hearts. What is the probability that the first card drawn was a spade?

5. Two people, whom we call South and North, draw 13 cards out of a deck of 52. South has two aces. What is the probability that North has (a) none? (b) One? (c) The other two?

6. An urn contains 8 red, 7 blue, and 5 green balls. You draw out two balls and they are different colors. Given this, what is the probability that the two balls were red and blue?

7. Suppose 60% of the people subscribe to newspaper A, 40% to newspaper B, and 30% to both. If we pick a person at random who subscribes to at least one newspaper, what is the probability that she subscribes to newspaper A?

8. In a town 40% of families have a dog and 30% have a cat. 25% of families with a dog also have a cat. (a) What fraction of people have a dog or cat? (b) What is the probability that a family with a cat has a dog?

9. Plumber Bob does 40% of the plumbing jobs in a small town. 30% of the people in town are unhappy with their plumbers, but 50% of Bob's customers are unhappy with his work. If your neighbor is not happy with his plumber, what is the probability that it was Bob?

10. An ectopic pregnancy is twice as likely if a woman smokes cigarettes. If 25% of women of childbearing age are smokers, what fraction of ectopic pregnancies occur to smokers?

11. Brown eyes are dominant over blue. That is, there are two alleles B and b. bb individuals have blue eyes but other combinations have brown eyes. Your parents and you have brown eyes but your brother has blue. So you can infer that both of your parents are heterozygotes, that is, have genetic type Bb. Given this information what is the probability that you are a homozygote.

12. Suppose that the probability that a married man votes is 0.45, the probability a married woman votes is 0.4, and the probability a woman votes given that her husband does is 0.6. What is the probability that (a) both vote? (b) A man votes given that his wife does?

13. Two events have $P(A) = 1/4$, $P(B|A) = 1/2$, and $P(A|B) = 1/3$. Compute $P(A \cap B)$, $P(B)$, and $P(A \cup B)$.

14. A, B, and C are events with $P(A) = 0.3$, $P(B) = 0.4$, and $P(C) = 0.5$, A and B are disjoint, A and C are independent, and $P(B|C) = 0.1$. Find $P(A \cup B \cup C)$.

Two-stage experiments

15. From a signpost that says MIAMI two letters fall off. A friendly drunk puts the two letters back into the two empty slots at random. What is the probability that the sign still says MIAMI?

16. Two balls are drawn from an urn with balls numbered from 1 up to 10. What is the probability that the two numbers will differ by more (>) than 3?

17. How can 5 black and 5 white balls be put into two urns to maximize the probability that a white ball is drawn when we draw from a randomly chosen urn?

18. Suppose we draw k cards out of a deck. What is the probability that we do not draw an ace? Is the answer larger or smaller than $(3/4)^k$?

19. You and a friend each roll two dice. What is the probability that you both have the same two numbers?

20. In a dice game the "dealer" rolls two dice, the player rolls two dice, and the player wins if his total is larger than the dealer's. What is the probability that the player wins?

21. What is the most likely total for the sum of four dice and what is its probability?

22. Charlie draws 5 cards out of a deck of 52. If he gets at least three of one suit, he discards the cards not of that suit and then draws until he again has 5 cards. For example, if he gets 3 hearts, 1 club, and 1 spade, he throws the 2 nonhearts away and draws 2 more. What is the probability that he will end up with 5 cards of the same suit?

23. Suppose 60% of the people in a town will get exposed to flu in the next month. If you are exposed and not inoculated then the probability of your getting the flu is 80%, but if you are inoculated that probability drops to 15%. Of two executives at Beta Company, one is inoculated and one is not. What is the probability at least one will not get the flu? Assume that the events that determine whether or not they get the flu are independent.

24. John takes the bus with probability 0.3 and the subway with probability 0.7. He is late 40% of the time when he takes the bus, but only 20% of the time when he takes the subway. What is the probability that he is late for work?

25. The population of Cyprus is 70% Greek and 30% Turkish. 20% of the Greeks and 10% of the Turks speak English. What fraction of the people of Cyprus speak English?

26. You are going to meet a friend at the airport. Your experience tells you that the plane is late 70% of the time when it rains, but is late only 20% of the time when it does not rain. The weather forecast that morning calls for a 40% chance of rain. What is the probability that the plane will be late?

27. Two boys have identical piggy banks. The older boy has 18 quarters and 12 dimes in his; the younger boy, 2 quarters and 8 dimes. One day the two banks get mixed up. You pick up a bank at random and shake it until a coin comes out. What is the probability that you get a quarter? Note that there are 20 quarters and 20 dimes in all.

28. Suppose that the number of children in a family has the following distribution:

Number of children	0	1	2	3	4
Probability	0.15	0.25	0.3	0.2	0.1

Assume that each child is independently a girl or a boy with probability 1/2 each. If a family is picked at random what is the chance it has exactly two girls.

29. A student is taking a multiple-choice test in which each question has four possible answers. She knows the answers to 50% of the questions, can narrow the choices down to two 30% of the time, and does not know anything about 20% of the questions. What is the probability that she will correctly answer a question chosen at random from the test?

30. A student is taking a multiple-choice test in which each question has four possible answers. She knows the answers to 5 of the questions, can narrow the choices down to 2 in 3 cases, and does not know anything about 2 of the questions. What is the probability that she will correctly answer (a) 10, (b) 9, (c) 8, (d) 7, (e) 6, and (f) 5 questions?

31. Two boys, Charlie and Doug, take turns rolling two dice with Charlie going first. If Charlie rolls a 6 before Doug rolls a 7 he wins. What is the probability that Charlie wins?

32. Three boys take turns shooting a basketball and have probabilities 0.2, 0.3, and 0.5 of scoring a basket. Compute the probabilities for each boy to get the first basket.

33. Change the second and third probabilities in the last problem so that each boy has an equal chance of winning.

Bayes' formula

34. 5% of men and 0.25% of women are color blind. Assuming that there are an equal number of men and women, what is the probability that a color-blind person is a man?

35. The alpha fetal protein test is meant to detect spina bifida in unborn babies, a condition that affects 1 out of 1,000 children who are born. The literature on the test indicates that 5% of the time a healthy baby will cause a positive reaction. We will assume that the test is positive 100% of the time when spina bifida is present. Your doctor has just told you that your alpha fetal protein test was positive. What is the probability that your baby has spina bifida?

36. Binary digits, that is, 0's and 1's, are sent down a noisy communications channel. They are received as sent with probability 0.9 but errors occur with probability 0.1. Assuming that 0's and 1's are equally likely, what is the probability that a 1 was sent given that we received a 1?

37. To improve the reliability of the channel described in the last example, we repeat each digit in the message three times. What is the probability that 111 was sent given that (a) we received 101? (b) We received 000?

38. Two hunters shoot at a deer, which is hit by exactly one bullet. If the first hunter hits his target with probability 0.3 and the second with probability 0.6, what is the probability that the second hunter killed the deer? The answer is not 2/3. Do you think the answer is larger or smaller?

39. A cab was involved in a hit-and-run accident at night. Two cab companies green and blue operate 85% and 15% of the cabs in the city respectively. A witness identified the cab as blue. However, in a test only 80% of witnesses were able to correctly identify the cab color. Given this what is the probability that the cab involved in the accident was blue?

40. A student goes to class on a snowy day with probability 0.4, but on a nonsnowy day attends with probability 0.7. Suppose that 20% of the days in February are snowy. What is the probability that it snowed on February 7 given that the student was in class on that day?

41. A company gave a test to 100 salesman, 80 with good sales records and 20 with poor sales records. 60% of the good salesman passed the test but only 30% of the poor salesmen did. Andy passed the test. Given this, what is the probability that he is a good salesman?

42. A company rates 80% of its employees as satisfactory and 20% as unsatisfactory. Personnel records indicate that 70% of the satisfactory workers had

prior experience but only 40% of the unsatisfactory workers did. If a person with previous work experience is hired, what is the probability that they will be a satisfactory worker?

43. A golfer hits his drive in the fairway with probability 0.7. When he hits his drive in the fairway he makes par 80% of the time. When he doesn't, he makes par only 30% of the time. He just made par on a hole. What is the probability that he hit his drive in the fairway?

44. You are about to have an interview for Harvard Law School. 60% of the interviewers are conservative and 40% are liberal. 50% of the conservatives smoke cigars but only 25% of the liberals do. Your interviewer lights up a cigar. What is the probability that he is a liberal?

45. Five pennies are sitting on a table. One is a trick coin that has heads on both sides, but the other four are normal. You pick up a penny at random and flip it four times, getting heads each time. Given this, what is the probability that you picked up the two-headed penny?

46. One slot machine pays off $1/2$ of the time, while another pays off $1/4$ of the time. We pick one of the machines and play it 6 times, winning 3 times. What is the probability we are playing the machine that pays off only $1/4$ of the time?

47. A student is taking a multiple-choice exam in which each question has four possible answers. She knows the answers to 60% of the questions and guesses at the others. What is the probability that she guessed given that she got question 12 right?

48. 20% of people are "accident-prone" and have a probability 0.15 of having an accident in a 1-year period in contrast to a probability of 0.05 for the other 80% of people. (a) If we pick a person at random, what is the probability that they will have an accident this year? (b) What is the probability a person is accident-prone if they had an accident last year? (c) What is the probability that they will have an accident this year if they had one last year?

49. One die has 4 red and 2 white sides; a second has 2 red and 4 white sides. (a) If we pick a die at random and roll it, what is the probability that the result is a red side? (b) If the first result is a red side and we roll the same die again, what is the probability of a second red side?

50. A particular football team is known to run 40% of its plays to the left and 60% to the right. When the play goes to the right, the right tackle shifts his stance 80% of the time, but does so only 10% of the time when the play goes to the left. As the team sets up for the play the right tackle shifts his stance. What is the probability that the play will go to the right?

51. A company gives a test to 100 salesmen, 80 with good sales records and 20 with poor records. 60% of the good salesmen pass the test, but only 30% of the poor salesmen do. A new applicant takes the test and passes. What is the probability that he is a good salesman?

52. You are a serious student who studies on Friday nights but your roommate goes out and has a good time. 40% of the time he goes out with his girlfriend; 60% of the time he goes to a bar. 30% of the times when he goes out with his girlfriend he spends the night at her apartment. 40% of the times when he goes to a bar he gets in a fight and gets thrown in jail. You wake up on Saturday morning and your roommate is not home. What is the probability that he is in jail?

53. Two masked robbers try to rob a crowded bank during the lunch hour but the teller presses a button that sets off an alarm and locks the front door. The robbers, realizing they are trapped, throw away their masks and disappear into the chaotic crowd. Confronted with 40 people claiming they are innocent, the police give everyone a lie detector test. Suppose that guilty people are detected with probability 0.95, and innocent people appear to be guilty with probability 0.01. What is the probability that Mr. Jones is guilty given that the lie detector says he is?

54. Three bags lie on the table. One has two gold coins, one has two silver coins, and one has one silver and one gold. You pick a bag at random, and pick out one coin. If this coin is gold, what is the probability that you picked from the bag with two gold coins?

55. In a certain city, 30% of the people are Conservatives, 50% are Liberals, and 20% are Independents. In a given election, 2/3 of the Conservatives voted, 80% of the Liberals voted, and 50% of the Independents voted. If we pick a voter at random what is the probability that she is Liberal?

56. An undergraduate student has asked a professor for a letter of recommendation. He estimates that the probability he will get the job is 0.8 with a strong letter, 0.4 with a medium letter, and 0.1 with a weak letter. He also believes that the probabilities that the letter will be strong, medium, or weak are 0.5, 0.3, and 0.2. What is the probability that the letter was strong given that he got the job.

57. A group of 20 people go out to dinner. 10 go to an Italian restaurant, 6 to a Japanese restaurant, and 4 to a French restaurant. The fractions of people satisfied with their meals were 4/5, 2/3, and 1/2 respectively. The next day the person you are talking to was satisfied with what they ate. What is the probability that they went to the Italian restaurant? The Japanese restaurant? The French restaurant?

58. 1 out of 1,000 births results in fraternal twins; 1 out of 1,500 births results in identical twins. Identical twins must be the same sex but the sexes of fraternal twins are independent. If two girls are twins, what is the probability that they are fraternal twins?

59. Consider the following data on traffic accidents:

Age group	% of drivers	Accident probability
16–25	15	0.10
26–45	35	0.04
46–65	35	0.06
>65	15	0.08

Calculate (a) the probability that a randomly chosen driver will have an accident this year, and (b) the probability that a driver is between 46 and 65 given that they had an accident.

Joint distributions

60. Suppose we draw two tickets from a hat that contains tickets numbered 1, 2, 3, 4. Let X be the first number drawn and Y be the second. Find the joint distribution of X and Y.

61. Suppose we roll one die repeatedly and let N_i be the number of the roll on which i first appears. Find the joint distribution of N_1 and N_6.

62. Compute (a) $P(X = 1|Y = 1)$ and (b) $P(X = 2|Y = 2)$ for the following joint distribution:

Y	$X = 1$	2	3
1	0.1	0.2	0.3
2	0.15	0.15	0
3	0.05	0	0.05

63. Compute (a) $P(X = 2|Y = 3)$ and (b) $P(Y = 3|X = 3)$ for the following joint distribution:

Y	$X = 1$	2	3
1	0.2	0.15	0.05
2	0.10	0	0.10
3	0.05	0.15	0.20

64. Using the clues given below, fill in the rest of the joint distribution. There is only one answer.

Y	$X = 0$	3	6
1	?	?	?
2	0.1	0.05	?

(a) $P(Y = 2|X = 0) = 1/4$, (b) X and Y are independent.

65. Using the clues given below, fill in the rest of the joint distribution. There is only one answer:

Y	$X = 1$	2	3
1	?	?	?
2	?	0	?
3	0	?	0

For $k = 1, 2, 3$, (a) $P(Y = 1|X = k) = 2/3$, (b) $P(X = k|Y = 1) = k/6$.

66. Fill in the rest of the joint distribution so that X and Y are independent. There are two possible answers:

Y	$X = 0$	1
0	?	2/9
1	2/9	?

4

Markov Chains

4.1 Definitions and examples

The importance of Markov chains comes from two facts: (i) there are a large number of physical, biological, economic, and social phenomena that can be described in this way, and (ii) there is a well-developed theory that allows us to do computations. We begin with a famous example, then describe the property that is the defining feature of Markov chains.

Example 4.1

Gambler's ruin. Consider a gambling game in which on any turn you win \$1 with probability $p = 0.4$ or lose \$1 with probability $1 - p = 0.6$. Suppose further that you adopt the rule that you quit playing if your fortune reaches \$$N$. Of course, if your fortune reaches \$0, the casino makes you stop.

Let X_n be the amount of money you have after n plays. I claim that your fortune, X_n, has the "Markov property." In words, this means that given the current state, any other information about the past is irrelevant for predicting the next state X_{n+1}. To check this, we note that if you are still playing at time n, that is, your fortune $X_n = i$ with $0 < i < N$, then for any possible history of your wealth $i_{n-1}, i_{n-2}, \dots, i_1, i_0$

$$P(X_{n+1} = i + 1 | X_n = i, X_{n-1} = i_{n-1}, \dots, X_0 = i_0) = 0.4$$

since to increase your wealth by 1 unit you have to win your next bet and the outcome of the previous bets has no useful information for predicting the next outcome.

Turning now to the formal definition, we say that X_n is a discrete-time **Markov chain** with **transition probability** $p(i, j)$ if for any $j, i, i_{n-1}, \dots, i_0$,

$$P(X_{n+1} = j | X_n = i, X_{n-1} = i_{n-1}, \dots, X_0 = i_0) = p(i, j) \tag{4.1}$$

Equation (4.1), also called the "Markov property," says that the conditional probability $X_{n+1} = j$ given the entire history $X_n = i, X_{n-1} = i_{n-1}, \dots, X_1 = i_1$,

$X_0 = i_0$ is the same as the conditional probability $X_{n+1} = j$ given only the previous state $X_n = i$. This is what we mean when we say that "given the current state any other information about the past is irrelevant for predicting X_{n+1}."

In formulating (4.1) we have restricted our attention to the **temporally homogeneous** case in which

$$p(i, j) = P(X_{n+1} = j | X_n = i)$$

does not depend on time n. Intuitively, the transition probability gives the rules of the game. It is the basic information needed to describe a Markov chain. In the case of the gambler's ruin chain, the transition probability has

$$p(i, i+1) = 0.4, \qquad p(i, i-1) = 0.6, \qquad \text{if } 0 < i < N$$
$$p(0, 0) = 1 \qquad p(N, N) = 1$$

When $N = 5$, the matrix is

	0	**1**	**2**	**3**	**4**	**5**
0	1.0	0	0	0	0	0
1	0.6	0	0.4	0	0	0
2	0	0.6	0	0.4	0	0
3	0	0	0.6	0	0.4	0
4	0	0	0	0.6	0	0.4
5	0	0	0	0	0	1.0

but picture is perhaps more informative:

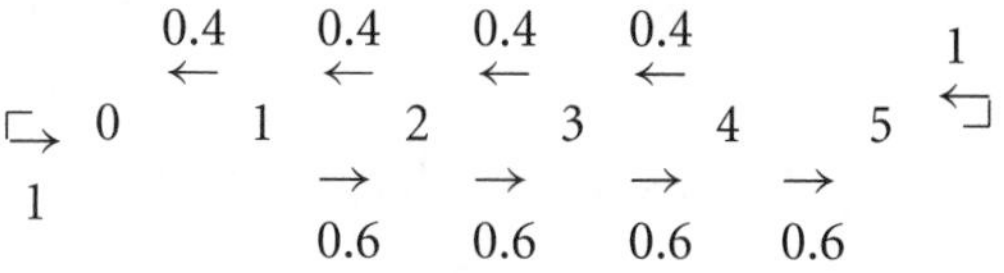

$$\begin{array}{cccccccc} & 0.4 & 0.4 & 0.4 & 0.4 & & & 1 \\ & \leftarrow & \leftarrow & \leftarrow & \leftarrow & & & \hookleftarrow \\ \hookrightarrow & 0 & 1 & 2 & 3 & 4 & 5 & \\ & & \rightarrow & \rightarrow & \rightarrow & \rightarrow & & \\ 1 & & 0.6 & 0.6 & 0.6 & 0.6 & & \end{array}$$

Example 4.2

Wright–Fisher model. We consider a fixed population of N genes that can be one of two types: A or a. These types are called alleles. In the simplest version of this model the population at time $n+1$ is obtained by drawing with replacement from the population at time n. In this case if we let X_n be the number of A alleles at time n, then X_n is a Markov chain with transition probability

$$p(i, j) = \binom{N}{j} \left(\frac{i}{N}\right)^j \left(1 - \frac{i}{N}\right)^{N-j} \qquad 0 \le i, j \le N$$

since the right-hand side is the binomial distribution for N independent trials with success probability i/N. Note that when $i = 0$, $p(0, 0) = 1$, and when $i = N$, $p(N, N) = 1$.

$A\ a\ a\ A$	
$A\ A\ a\ a$	
$A\ a\ A\ a$	
$a\ A\ A\ A$	
$A\ a\ A\ A$	
Time n	Time $n+1$

In the gambler's ruin chain and the Wright–Fisher model the states 0 and N are **absorbing states**. Once we enter these states we can never leave. The long-run behavior of these models is not very interesting; they will eventually enter one of the absorbing states and stay there forever. To make the Wright–Fisher model more interesting and more realistic, we can introduce the possibility of mutations: an A that is drawn ends up being an a in the next generation with probability u, while an a that is drawn ends up being an A in the next generation with probability v. In this case the probability an A is produced by a given draw is

$$\rho_i = \frac{i}{N}(1-u) + \frac{N-i}{N}v$$

that is, we can get an A by drawing an A and not having a mutation or by drawing an a and having a mutation. Since the draws are independent the transition probability still has the binomial form

$$p(i, j) = \binom{N}{j}(\rho_i)^j(1-\rho_i)^{N-j} \tag{4.2}$$

Moving from biology to physics

Example 4.3 **Ehrenfest chain.** We imagine two cubical volumes of air connected by a small hole. In the mathematical version, we have two "urns," that is, two of the exalted trash cans of probability theory, in which there are a total of N balls. We pick one of the N balls at random and move it to the other urn.

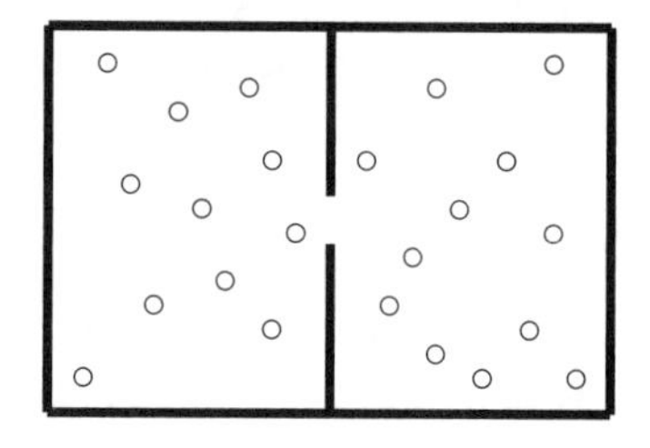

Let X_n be the number of balls in the "left" urn after the nth draw. It should be clear that X_n has the Markov property; that is, if we want to guess the state at time $n+1$, then the current number of balls in the left urn, X_n, is the only relevant information from the observed sequence of states $X_n, X_{n-1}, \ldots, X_1, X_0$. To check this we note that

$$P(X_{n+1} = i+1 | X_n = i, X_{n-1} = i_{n-1}, \ldots, X_0 = i_0) = \frac{N-i}{N}$$

since to increase the number we have to pick one of the $N-i$ balls in the other urn. The number can also decrease by 1 with probability i/N. In symbols, we have computed that the transition probability is given by

$$p(i, i+1) = \frac{N-i}{N}, \qquad p(i, i-1) = \frac{i}{N} \quad \text{for } 0 \le i \le N$$

with $p(i, j) = 0$ otherwise. When $N = 5$, for example, the matrix is

	0	**1**	**2**	**3**	**4**	**5**
0	0	5/5	0	0	0	0
1	1/5	0	4/5	0	0	0
2	0	2/5	0	3/5	0	0
3	0	0	3/5	0	2/5	0
4	0	0	0	4/5	0	1/5
5	0	0	0	0	5/5	0

Here, we have written 1 as 5/5 to emphasize the pattern in the diagonals of the matrix.

Moving from science to business

Example 4.4

Inventory chain. An electronics store sells a video game system. If at the end of the day, the number of units they have on hand is 1 or 0, they order enough new units, so their total on hand is 5. This chain is an example of an s, S inventory control policy with $s = 1$ and $S = 5$. That is, when the stock on hand falls to s or below, we order enough to bring it back up to S.

For simplicity we assume that the new merchandise arrives before the store opens the next day. Let X_n be the number of units on hand at the end of the nth day. If we assume that the number of customers who want to buy a video game system each day is 0, 1, 2, or 3 with probabilities 0.3, 0.4, 0.2, and 0.1, then we have the following transition matrix:

	0	**1**	**2**	**3**	**4**	**5**
0	0	0	0.1	0.2	0.4	0.3
1	0	0	0.1	0.2	0.4	0.3
2	0.3	0.4	0.3	0	0	0
3	0.1	0.2	0.4	0.3	0	0
4	0	0.1	0.2	0.4	0.3	0
5	0	0	0.1	0.2	0.4	0.3

To explain the entries we note that if $X_n \geq 2$, then no ordering is done, so what we have at the end of the day is the supply minus the demand. If $X_n = 2$ and the demand is 3 or more, or if $X_n = 3$ and the demand is 4, we end up with 0 units at the end of the day and at least one unhappy customer. If $X_n = 0$ or 1, then we will order enough so that at the beginning of the day we have 5, so the result at the end of the day is the same as if $X_n = 5$.

Markov chains are described by giving their transition probabilities. To create a chain, we can write down any $n \times n$ matrix, provided that the entries satisfy:

(i) $p(i, j) \geq 0$, since they are probabilities.
(ii) $\sum_j p(i, j) = 1$, since when $X_n = i$, X_{n+1} will be in some state j.

The equation in (ii) is read "sum $p(i, j)$ over all possible values of j." In words, the last two conditions say, the entries of the matrix are nonnegative and each **row** of the matrix sums to 1.

Any matrix with properties (i) and (ii) gives rise to a Markov chain, X_n. To construct the chain we can think of playing a board game. When we are in state i, we roll a die (or generate a random number on a computer) to pick the next state, going to j with probability $p(i, j)$. To illustrate this we now introduce some simple examples.

Example 4.5

Weather chain. Let X_n be the weather on day n in Ithaca, NY, which we assume is either 1 = rainy or 2 = sunny. Even though the weather is not exactly a Markov chain, we can propose a Markov chain model for the weather by writing down a transition probability:

	1	**2**
1	0.6	0.4
2	0.2	0.8

The table says, for example, the probability that a rainy day (state 1) is followed by a sunny day (state 2) is $p(1, 2) = 0.4$.

Example 4.6

Social mobility. Let X_n be a family's social class in the nth generation, which we assume is either 1 = lower, 2 = middle, or 3 = upper. In our simple version of sociology, changes of status are a Markov chain with the following transition probability:

	1	2	3
1	0.7	0.2	0.1
2	0.3	0.5	0.2
3	0.2	0.4	0.4

Example 4.7

Brand preference. Suppose there are three types of laundry detergent, 1, 2, and 3, and let X_n be the brand chosen on the nth purchase. Customers who try these brands are satisfied and choose the same thing again with probabilities 0.8, 0.6, and 0.4 respectively. When they change they pick one of the other two brands at random. The transition probability is

	1	2	3
1	0.8	0.1	0.1
2	0.2	0.6	0.2
3	0.3	0.3	0.4

Example 4.8

Two-stage Markov chains. In a Markov chain the distribution of X_{n+1} depends only on X_n. This can easily be generalized to case in which the distribution of X_{n+1} depends only on (X_n, X_{n-1}). For a concrete example consider a basketball player who makes a shot with the following probabilities:

1/2 if he has missed the last two times
2/3 if he has hit one of his last two shots
3/4 if he has hit both of his last two shots

To formulate a Markov chain to model his shooting, we let the states of the process be the outcomes of his last two shots: $\{HH, HM, MH, MM\}$, where M is short for miss and H for hit. The transition probability is

	HH	HM	MH	MM
HH	3/4	1/4	0	0
HM	0	0	2/3	1/3
MH	2/3	1/3	0	0
MM	0	0	1/2	1/2

To explain suppose the state is HM; that is, $X_{n-1} = H$ and $X_n = M$. In this case the next outcome will be H with probability 2/3. When this occurs the next state will be $(X_n, X_{n+1}) = (M, H)$ with probability 2/3. If he misses an event of probability 1/3, $(X_n, X_{n+1}) = (M, M)$.

The hot hand is a phenomenon known to everyone who plays or watches basketball. After making a couple of shots, players are thought to "get into a groove" so that subsequent successes are more likely. Purvis Short of the Golden State Warriors describes this more poetically as

> you're in a world all your own. It's hard to describe. But the basket seems to be so wide. No matter what you do, you know the ball is going to go in.

Unfortunately for basketball players, data collected by Tversky and Gliovich (*Chance*, Vol. 2 (1989), No. 1, pp. 16–21) show that this is a misconception. The next table gives data for the conditional probability of hitting a shot after missing the last three, missing the last two, . . . , hitting the last three, for nine players of the Philadelphia 76ers: Darryl Dawkins (403), Maurice Cheeks (339), Steve Mix (351), Bobby Jones (433), Clint Richardson (248), Julius Erving (884), Andrew Toney (451), Caldwell Jones (272), and Lionel Hollins (419). The numbers in parentheses are the number of shots for each player.

$P(H\|3M)$	$P(H\|2M)$	$P(H\|1M)$	$P(H\|1H)$	$P(H\|2H)$	$P(H\|3H)$
0.88	0.73	0.71	0.57	0.58	0.51
0.77	0.60	0.60	0.55	0.54	0.59
0.70	0.56	0.52	0.51	0.48	0.36
0.61	0.58	0.58	0.53	0.47	0.53
0.52	0.51	0.51	0.53	0.52	0.48
0.50	0.47	0.56	0.49	0.50	0.48
0.50	0.48	0.47	0.45	0.43	0.27
0.52	0.53	0.51	0.43	0.40	0.34
0.50	0.49	0.46	0.46	0.46	0.32

In fact, the data support the opposite assertion: after missing a player is more conservative about the shots they take and will hit more frequently.

4.2 Multistep transition probabilities

The previous section introduced several examples to think about. The basic question concerning Markov chains is what happens in the long run? In the case of the weather chain, does the probability that day n is sunny converge to a limit? In the case of the social mobility and brand preference chain, do the fractions of the population in the three income classes (or that buy each of the three types of detergent) stabilize as time goes on? The first step in answering these questions is to figure out what happens in the Markov chain after two or more steps.

The transition probability $p(i, j) = P(X_{n+1} = j | X_n = i)$ gives the probability of going from i to j in one step. Our goal in this section is to compute the probability of going from i to j in $m > 1$ steps:

$$p^m(i, j) = P(X_{n+m} = j | X_n = i)$$

For a concrete example, we start with the transition probability of the social mobility chain:

	1	**2**	**3**
1	0.7	0.2	0.1
2	0.3	0.5	0.2
3	0.2	0.4	0.4

To warm up we consider

Example 4.9

Suppose the family starts in the middle class (state 2) in generation 0. What is the probability that the generation 1 rises to the upper class (state 3) and generation 2 falls to the lower class (state 1)?

Intuitively, the Markov property implies that starting from state 2 the probability of jumping to 1 and then to 3 is given by $p(2, 3)p(3, 1)$. To get this conclusion from the definitions, we note that using the definition of conditional probability,

$$P(X_2 = 1, X_1 = 3 | X_0 = 2) = \frac{P(X_2 = 1, X_1 = 3, X_0 = 2)}{P(X_0 = 2)}$$

Multiplying and dividing by $P(X_1 = 3, X_0 = 2)$,

$$= \frac{P(X_2 = 1, X_1 = 3, X_0 = 2)}{P(X_1 = 3, X_0 = 2)} \cdot \frac{P(X_1 = 3, X_0 = 2)}{P(X_0 = 2)}$$

Using the definition of conditional probability,

$$= P(X_2 = 1 | X_1 = 3, X_0 = 2) \cdot P(X_1 = 3 | X_0 = 2)$$

By the Markov property (4.1), the last expression is

$$P(X_2 = 1 | X_1 = 3) \cdot P(X_1 = 3 | X_0 = 2) = p(2, 3)p(3, 1)$$ □

Moving on to the real question

Example 4.10

Suppose the family starts in the middle class (state 2) in generation 0. What is the probability that generation 2 will be in the lower class (state 1)?

To do this we simply have to consider the three possible states for generation 1 and use the previous computation.

$$P(X_2 = 1|X_0 = 2) = \sum_{k=1}^{3} P(X_2 = 1, X_1 = k|X_0 = 2) = \sum_{k=1}^{3} p(2, k)p(k, 1)$$
$$= (0.3)(0.7) + (0.5)(0.3) + (0.2)(0.2)$$
$$= 0.21 + 0.15 + 0.04 = 0.40$$

There is nothing special here about the states 2 and 1 here. By the same reasoning,

$$P(X_2 = j|X_0 = i) = \sum_{k=1}^{3} p(i, k)\, p(k, j)$$

The right-hand side of the last equation gives the (i, j)th entry of the matrix p multiplied by itself.

To explain this, we note that to compute $p^2(2, 1)$ we multiplied the entries of the second row by those in the first column:

$$\begin{pmatrix} . & . & . \\ 0.3 & 0.5 & 0.2 \\ . & . & . \end{pmatrix} \begin{pmatrix} 0.7 & . & . \\ 0.3 & . & . \\ 0.2 & . & . \end{pmatrix} = \begin{pmatrix} . & . & . \\ 0.40 & . & . \\ . & . & . \end{pmatrix}$$

If we wanted $p^2(1, 3)$, we would multiply the first row by the third column:

$$\begin{pmatrix} 0.7 & 0.2 & 0.1 \\ . & . & . \\ . & . & . \end{pmatrix} \begin{pmatrix} . & . & 0.1 \\ . & . & 0.2 \\ . & . & 0.4 \end{pmatrix} = \begin{pmatrix} . & . & 0.15 \\ . & . & . \\ . & . & . \end{pmatrix}$$

When all of the computations are done, we have

$$\begin{pmatrix} 0.7 & 0.2 & 0.1 \\ 0.3 & 0.5 & 0.2 \\ 0.2 & 0.4 & 0.4 \end{pmatrix} \begin{pmatrix} 0.7 & 0.2 & 0.1 \\ 0.3 & 0.5 & 0.2 \\ 0.2 & 0.4 & 0.4 \end{pmatrix} = \begin{pmatrix} 0.57 & 0.28 & 0.15 \\ 0.40 & 0.39 & 0.21 \\ 0.34 & 0.40 & 0.26 \end{pmatrix}$$

The two-step transition probability $p^2 = p \cdot p$. Based on this you can probably leap to the next conclusion:

Theorem 4.1. *The m-step transition probability*

$$p^m(i, j) = P(X_{n+m} = j|X_n = i) \tag{4.3}$$

is the mth power of the transition matrix p, that is, $p \cdot p \cdot \cdots \cdot p$, where there are m terms in the product.

The key ingredient in proving this is

Chapman–Kolmogorov equation:

$$p^{m+n}(i, j) = \sum_k p^m(i, k)\, p^n(k, j) \tag{4.4}$$

Once this is proved, (4.3) follows, since taking $n = 1$ in (4.4), we see that $p^{m+1} = p^m \cdot p$.

Why is (4.4) true? To go from i to j in $m + n$ steps, we have to go from i to some state k in m steps and then from k to j in n steps. The Markov property implies that the two parts of our journey are independent. □

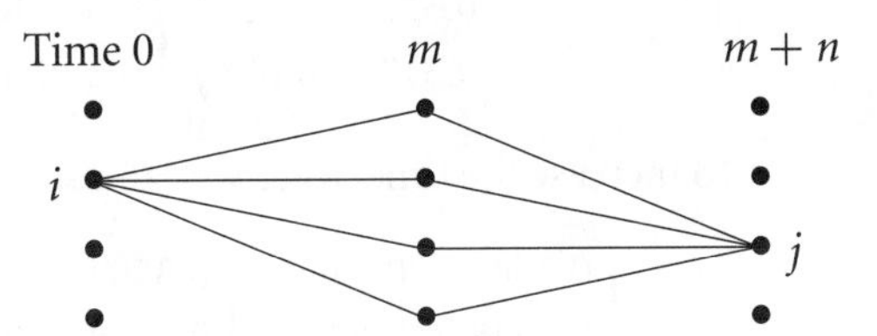

Proof of (4.4). The *independence* in the second sentence of the previous explanation is the mysterious part. To show this, we combine Examples 4.9 and 4.10. Breaking things down according to the state at time m,

$$P(X_{m+n} = j | X_0 = i) = \sum_k P(X_{m+n} = j, X_m = k | X_0 = i)$$

Repeating the computation in Example 4.9, the definition of conditional probability implies

$$P(X_{m+n} = j, X_m = k | X_0 = i) = \frac{P(X_{m+n} = j, X_m = k, X_0 = i)}{P(X_0 = i)}$$

Multiplying and dividing by $P(X_m = k, X_0 = i)$ gives

$$= \frac{P(X_{m+n} = j, X_m = k, X_0 = i)}{P(X_m = k, X_0 = i)} \cdot \frac{P(X_m = k, X_0 = i)}{P(X_0 = i)}$$

Using the definition of conditional probability we have

$$= P(X_{m+n} = j | X_m = k, X_0 = i) \cdot P(X_m = k | X_0 = i)$$

By the Markov property (4.1), the last expression is

$$= P(X_{m+n} = j | X_m = k) \cdot P(X_m = k | X_0 = i) = p^m(i, k)\, p^n(k, j)$$

and we have proved (4.4). □

Having established (4.4), we now return to computations. We begin with the weather chain

$$\begin{pmatrix} 0.6 & 0.4 \\ 0.2 & 0.8 \end{pmatrix} \begin{pmatrix} 0.6 & 0.4 \\ 0.2 & 0.8 \end{pmatrix} = \begin{pmatrix} 0.44 & 0.56 \\ 0.28 & 0.72 \end{pmatrix}$$

Multiplying again $p^2 \cdot p = p^3$

$$\begin{pmatrix} 0.44 & 0.56 \\ 0.28 & 0.72 \end{pmatrix} \begin{pmatrix} 0.6 & 0.4 \\ 0.2 & 0.8 \end{pmatrix} = \begin{pmatrix} 0.376 & 0.624 \\ 0.312 & 0.688 \end{pmatrix}$$

and then $p^3 \cdot p = p^4$

$$\begin{pmatrix} 0.376 & 0.624 \\ 0.312 & 0.688 \end{pmatrix} \begin{pmatrix} 0.6 & 0.4 \\ 0.2 & 0.8 \end{pmatrix} = \begin{pmatrix} 0.3504 & 0.6496 \\ 0.3248 & 0.6752 \end{pmatrix}$$

To increase the time faster we can use (4.4) to conclude that $p^4 \cdot p^4 = p^8$:

$$\begin{pmatrix} 0.3504 & 0.6496 \\ 0.3248 & 0.6752 \end{pmatrix} \begin{pmatrix} 0.3504 & 0.6496 \\ 0.3248 & 0.6752 \end{pmatrix} = \begin{pmatrix} 0.33377 & 0.66623 \\ 0.33311 & 0.66689 \end{pmatrix}$$

Multiplying again $p^8 \cdot p^8 = p^{16}$

$$\begin{pmatrix} 0.33333361 & 0.66666689 \\ 0.33333319 & 0.66666681 \end{pmatrix}$$

Based on the last calculation, one might guess that as n gets large the matrix becomes closer and closer to

$$\begin{pmatrix} 1/3 & 2/3 \\ 1/3 & 2/3 \end{pmatrix}$$

This is true and will be explained in the next section.

4.3 Stationary distributions

Our first step is to consider

What happens when the initial state is random? Breaking things down according to the value of the initial state and using the definition of conditional probability,

$$\begin{aligned} P(X_n = j) &= \sum_i P(X_0 = i, X_n = j) \\ &= \sum_i P(X_0 = i) P(X_n = j | X_0 = i) \end{aligned}$$

If we introduce $q(i) = P(X_0 = i)$, then the last equation can be written as

$$P(X_n = j) = \sum_i q(i) p^n(i, j) \tag{4.5}$$

In words, we multiply the transition matrix on the left by the vector q of initial probabilities. If there are k states, then $p^n(x, y)$ is a $k \times k$ matrix. So to make the matrix multiplication work out right, we should take q as a $1 \times k$ matrix or a "row vector."

For a concrete example consider the weather chain (Example 4.5) and suppose that the initial distribution is $q(1) = 0.3$ and $q(2) = 0.7$. In this case

$$\begin{pmatrix} 0.3 & 0.7 \end{pmatrix} \begin{pmatrix} 0.6 & 0.4 \\ 0.2 & 0.8 \end{pmatrix} = \begin{pmatrix} 0.32 & 0.68 \end{pmatrix}$$

since

$$0.3(0.6) + 0.7(0.2) = 0.32$$

$$0.3(0.4) + 0.7(0.8) = 0.68$$

For a second example consider the social mobility chain (Example 4.6) and suppose that the initial distribution $q(1) = 0.5$, $q(2) = 0.2$, and $q(3) = 0.3$. Multiplying the vector q by the transition probability gives the vector of probabilities at time 1.

$$\begin{pmatrix} 0.5 & 0.2 & 0.3 \end{pmatrix} \begin{pmatrix} 0.7 & 0.2 & 0.1 \\ 0.3 & 0.5 & 0.2 \\ 0.2 & 0.4 & 0.4 \end{pmatrix} = \begin{pmatrix} 0.47 & 0.32 & 0.21 \end{pmatrix}$$

To check the arithmetic note that the three entries on the right-hand side are

$$0.5(0.7) + 0.2(0.3) + 0.3(0.2) = 0.35 + 0.06 + 0.06 = 0.47$$

$$0.5(0.2) + 0.2(0.5) + 0.3(0.4) = 0.10 + 0.10 + 0.12 = 0.32$$

$$0.5(0.1) + 0.2(0.2) + 0.3(0.4) = 0.05 + 0.04 + 0.12 = 0.21$$

If the distribution at time 0 is the same as the distribution at time 1, then by the Markov property it will be the distribution at all times $n \geq 1$. Because of this q is called a **stationary distribution**. Stationary distributions have a special importance in the theory of Markov chains, so we will use a special letter π to denote solutions of the equation

$$\pi \cdot p = \pi$$

To have a mental picture of what happens to the distribution of probability when one step of the Markov chain is taken, it is useful to think that we have

$q(i)$ pounds of sand at state i, with the total amount of sand $\sum_i q(i)$ being 1 pound. When a step is taken in the Markov chain, a fraction $p(i, j)$ of the sand at i is moved to j. The distribution of sand when this has been done is

$$q \cdot p = \sum_i q(i) p(i, j)$$

If the distribution of sand is not changed by this procedure, q is a stationary distribution.

General two-state transition probability.

	1	2
1	$1 - a$	a
2	b	$1 - b$

We have written the chain in this way, so the stationary distribution has a simple formula

$$\pi(1) = \frac{b}{a+b} \qquad \pi(2) = \frac{a}{a+b} \tag{4.6}$$

As a first check on this formula we note that in the weather chain, $a = 0.4$ and $b = 0.2$ which gives $(1/3, 2/3)$ as we found before. We can prove this work in general by drawing a picture:

$$\frac{b}{a+b} \quad \overset{1}{\bullet} \underset{b}{\overset{a}{\rightleftarrows}} \overset{2}{\bullet} \quad \frac{a}{a+b}$$

In words, the amount of sand that flows from 1 to 2 is the same as the amount that flows from 2 to 1, so the amount of sand at each site stays constant. To check algebraically that this is true:

$$\begin{aligned} \frac{b}{a+b}(1-a) + \frac{a}{a+b}b &= \frac{b - ba + ab}{a+b} = \frac{b}{a+b} \\ \frac{b}{a+b}a + \frac{a}{a+b}(1-b) &= \frac{ba + a - ab}{a+b} = \frac{a}{a+b} \end{aligned} \tag{4.7}$$

Formula (4.6) gives the stationary distribution for any two-state chain, so we progress now to the three-state case and consider the brand preference chain (Example 4.7). The equation $\pi p = \pi$ says

$$\begin{pmatrix} \pi_1 & \pi_2 & \pi_3 \end{pmatrix} \begin{pmatrix} 0.8 & 0.1 & 0.1 \\ 0.2 & 0.6 & 0.2 \\ 0.3 & 0.3 & 0.4 \end{pmatrix} = \begin{pmatrix} \pi_1 & \pi_2 & \pi_3 \end{pmatrix}$$

which translates into three equations:

$$
\begin{aligned}
0.8\pi_1 + 0.2\pi_2 + 0.3\pi_3 &= \pi_1 \\
0.1\pi_1 + 0.6\pi_2 + 0.3\pi_3 &= \pi_2 \\
0.1\pi_1 + 0.2\pi_2 + 0.4\pi_3 &= \pi_3
\end{aligned}
$$

Note that the columns of the matrix give the numbers in the rows of the equations. The third equation is redundant since if we add up the three equations, we get

$$\pi_1 + \pi_2 + \pi_3 = \pi_1 + \pi_2 + \pi_3$$

If we replace the third equation by $\pi_1 + \pi_2 + \pi_3 = 1$ and subtract π_1 from each side of the first equation and π_2 from each side of the second equation, we get

$$
\begin{aligned}
-0.2\pi_1 + 0.2\pi_2 + 0.3\pi_3 &= 0 \\
0.1\pi_1 - 0.4\pi_2 + 0.3\pi_3 &= 0 \\
\pi_1 + \pi_2 + \pi_3 &= 1
\end{aligned}
\tag{4.8}
$$

At this point we can solve the equations by hand or using a calculator.

By hand. We note that the third equation implies $\pi_3 = 1 - \pi_1 - \pi_2$ and substituting this in the first two gives

$$
\begin{aligned}
0.3 &= 0.5\pi_1 + 0.1\pi_2 \\
0.3 &= 0.2\pi_1 + 0.7\pi_2
\end{aligned}
$$

Multiplying the first equation by 0.7 and adding -0.1 times the second gives

$$1.8 = (0.35 - 0.02)\pi_1 \qquad \text{or} \qquad \pi_1 = 18/33 = 6/11$$

Multiplying the first equation by 0.2 and adding -0.5 times the second gives

$$-0.09 = (0.02 - 0.35)\pi_2 \qquad \text{or} \qquad \pi_2 = 9/33 = 3/11$$

Since the three probabilities add up to 1, $\pi_3 = 2/11$.

Using the TI-83 calculator is easier. To begin we write (4.8) in matrix form as

$$
\begin{pmatrix} \pi_1 & \pi_2 & \pi_3 \end{pmatrix}
\begin{pmatrix} -0.2 & 0.1 & 1 \\ 0.2 & -0.4 & 1 \\ 0.3 & 0.3 & 1 \end{pmatrix}
= \begin{pmatrix} 0 & 0 & 1 \end{pmatrix}
$$

If we let A be the 3×3 matrix in the middle this can be written as $\pi A = (0, 0, 1)$. Multiplying on each side by A^{-1}, we see that

$$\pi = (0, 0, 1) A^{-1}$$

which is the third row of A^{-1}. To compute A^{-1}, we enter A into our calculator (using the MATRX menu and the EDIT submenu), use the MATRX menu to put $[A]$ on the computation line, and then press the x^{-1}. Reading the third row we find that the stationary distribution is

$$(0.545454,\ 0.272727,\ 0.181818)$$

Converting the answer to fractions using the first entry in the MATH menu gives

$$(6/11,\ 3/11,\ 2/11)$$

Example 4.11 **Social mobility (continuation of 4.6)**

	1	2	3
1	0.7	0.2	0.1
2	0.3	0.5	0.2
3	0.2	0.4	0.4

Using the first two equations and the fact that the sum of the π's is 1,

$$\begin{aligned} 0.7\pi_1 + 0.3\pi_2 + 0.2\pi_3 &= \pi_1 \\ 0.2\pi_1 + 0.5\pi_2 + 0.4\pi_3 &= \pi_2 \\ \pi_1 + \pi_2 + \pi_3 &= 1 \end{aligned}$$

This translates into $\pi A = (0, 0, 1)$ with

$$A = \begin{pmatrix} -0.3 & 0.2 & 1 \\ 0.3 & -0.5 & 1 \\ 0.2 & 0.4 & 1 \end{pmatrix}$$

Note that here and in the previous example the first two columns of A consist of the first two columns of the transition probability with 1 subtracted from the diagonal entries, and the final column is all 1's. Computing the inverse and reading the last row gives

$$(0.468085,\ 0.340425,\ 0.191489)$$

Converting the answer to fractions using the first entry in the math menu gives

$$(22/47,\ 16/47,\ 9/47)$$

Example 4.12 **Basketball (continuation of 4.8)**

To find the stationary matrix in this case we can follow the same procedure. A consists of the first three columns of the transition matrix with 1 subtracted from the diagonal, and a final column of all 1's.

$$
\begin{array}{cccc}
-1/4 & 1/4 & 0 & 1 \\
0 & -1 & 2/3 & 1 \\
2/3 & 1/3 & -1 & 1 \\
0 & 0 & 1/2 & 1
\end{array}
$$

The answer is given by the fourth row of A^{-1}:

$$(0.5,\ 0.1875,\ 0.1875,\ 0.125) = (1/2,\ 3/16,\ 3/16,\ 1/8)$$

Thus the long-run fraction of time the player hits a shot is

$$\pi(HH) + \pi(MH) = 0.6875 = 11/36$$

Example 4.13 **Inventory chain (continuation of 4.4)**

As in the two previous examples, A consists of the first five columns of the transition matrix with 1 subtracted from the diagonal, and a final column of all 1s.

$$
\begin{array}{cccccc}
-1 & 0 & 0.1 & 0.2 & 0.4 & 1 \\
0 & -1 & 0.1 & 0.2 & 0.4 & 1 \\
0.3 & 0.4 & -0.7 & 0 & 0 & 1 \\
0.1 & 0.2 & 0.4 & -0.7 & 0 & 1 \\
0 & 0.1 & 0.2 & 0.4 & -0.7 & 1 \\
0 & 0 & 0.1 & 0.2 & 0.4 & 1
\end{array}
$$

The answer is given by the sixth row of A^{-1}:

$$(0.090862,\ 0.155646,\ 0.231006,\ 0.215605,\ 0.201232,\ 0.105646)$$

Converting the answer to fractions using the first entry in the math menu gives

$$(177/1{,}948,\ 379/2{,}435,\ 225/974,\ 105/487,\ 98/487,\ 1{,}029/9{,}740)$$

but the decimal version is probably more informative.

Example 4.14 **Ehrenfest chain (continuation of 4.3)**

Consider first the case $N = 5$. As in the three previous examples, A consists of the first five columns of the transition matrix with 1 subtracted from the diagonal, and a final column of all 1's.

$$\begin{matrix} -1 & 1 & 0 & 0 & 0 & 1 \\ 0.2 & -1 & 0.8 & 0 & 0 & 1 \\ 0 & 0.4 & -1 & 0.6 & 0 & 1 \\ 0 & 0 & 0.6 & -1 & 0.4 & 1 \\ 0 & 0 & 0 & 0.8 & -1 & 1 \\ 0 & 0 & 0 & 0 & 1 & 1 \end{matrix}$$

The answer is given by the sixth row of A^{-1}:

$$(0.03125,\ 0.15625,\ 0.3125,\ 0.3125,\ 0.15625,\ 0.03125)$$

Even without a calculator we can recognize these as

$$(1/32,\ 5/32,\ 10/32,\ 10/32,\ 5/32,\ 1/32)$$

the probabilities of 0–5 heads when we flip five coins.

Based on this we can guess that in general

$$\pi(k) = \binom{N}{k} / 2^N$$

Proof by computation. For $0 < k < N$, we can end up in state k only by coming up from $k-1$ or down from $k+1$, so

$$\begin{aligned} &\pi(k-1)p(k-1,k) + \pi(k+1)p(k+1,k) \\ &= \frac{\binom{N}{k-1}}{2^N} \cdot \frac{N-k+1}{N} + \frac{\binom{N}{k+1}}{2^N} \cdot \frac{k+1}{N} \\ &= \frac{1}{2^N}\left(\frac{(N-1)!}{(k-1)!(N-k)!} + \frac{(N-1)!}{(k)!(N-k+1)!}\right) \\ &= \frac{1}{2^N}\binom{N}{k}\left(\frac{k}{N} + \frac{N-k}{N}\right) = \pi(k) \end{aligned}$$

The only way to end up at 0 is by coming down from 1, so

$$\pi(1)p(1,0) = \frac{N}{2^N} \cdot \frac{1}{N} = \pi(0)$$

Similarly, the only way to end up at N is by coming up from $N-1$, so

$$\pi(N-1)p(N-1,N) = \frac{N}{2^N} \cdot \frac{1}{N} = \pi(0)$$

□

Proof by thinking. To determine the initial state, (a) flip N coins, with heads = in the left urn and tails = in the right. A transition of the chain corresponds to (b) picking a coin at random and turning it over. It is clear that the end result of (a) and (b) has all 2^N outcomes equally likely, so the state at time 1 is the same as the state at time 0. □

4.4 Limit behavior

In this section we give conditions that guarantee that as n gets large, $p^n(i, j)$ approaches its stationary distribution. We begin with the convergence in the two-state case.

Theorem 4.2. *Let p_n be the probability of being in state 1 after n steps. For a two-state Markov chain with transition probability*

$$\begin{matrix} 1-a & a \\ b & 1-b \end{matrix}$$

where $0 < a + b < 2$, we have

$$\left| p_n - \frac{b}{a+b} \right| = \left| p_0 - \frac{b}{a+b} \right| \cdot |1 - a - b|^n \tag{4.9}$$

In words, the transition probability converges to equilibrium exponentially fast. In the case of the weather chain, $|1 - a - b| = 0.4$, so the difference between p_n and the limit $b/(a + b)$ goes to 0 faster than $(0.4)^n$.

Proof. Using the Markov property we have for any $n \geq 1$ that

$$p_n = p_{n-1}(1 - a) + (1 - p_{n-1})b$$

In words, the chain is in state 1 at time n if it was in state 1 at time $n - 1$ (with probability p_{n-1}) and stays there (with probability $1 - a$), or if it was in state 2 (with probability $1 - p_{n-1}$) and jumps from 2 to 1 (with probability b). Since the probability of being in state 1 is constant when we start in the stationary distribution, see the first equation in (4.7):

$$\frac{b}{a+b} = \frac{b}{a+b}(1 - a) + \left(1 - \frac{b}{a+b}\right) b$$

Subtracting this equation from the one for p_n, we have

$$\begin{aligned} p_n - \frac{b}{a+b} &= \left(p_{n-1} - \frac{b}{a+b} \right)(1 - a) + \left(\frac{b}{a+b} - p_{n-1} \right) b \\ &= \left(p_{n-1} - \frac{b}{a+b} \right)(1 - a - b) \end{aligned}$$

If $0 < a+b < 2$, then $|1-a-b| < 1$ and we have

$$\left| p_n - \frac{b}{a+b} \right| = \left| p_{n-1} - \frac{b}{a+b} \right| \cdot |1-a-b|$$

In words, the difference $|p_n - b/(a+b)|$ will shrink by a factor $|1-a-b|$ at each step. Iterating the last equation gives the desired result. □

There are two cases $a = b = 0$ and $a = b = 1$ in which $p^n(i, j)$ does not converge to $\pi(i)$. In the first case the matrix is

$$\begin{pmatrix} 1 & 0 \\ 0 & 1 \end{pmatrix}$$

so the state never changes. This is called the **identity matrix** and denoted by I, since for any 2×2 matrix m, $I \cdot m = m$ and $m \cdot I = m$. In the second case the matrix is

$$p = \begin{pmatrix} 0 & 1 \\ 1 & 0 \end{pmatrix}$$

so the chain always jumps. In this case, $p^2 = I$, $p^3 = p$, $p^4 = I$, etc. To see that something similar can happen in a "real example," we consider

Example 4.15 **Ehrenfest chain.** Consider the chain defined in Example 4.3, and for simplicity suppose there are three balls. In this case the transition probability is

	0	**1**	**2**	**3**
0	0	3/3	0	0
1	1/3	0	2/3	0
2	0	2/3	0	1/3
3	0	0	3/3	0

In the second power of p, the zero pattern is shifted:

	0	**1**	**2**	**3**
0	1/3	0	2/3	0
1	0	7/9	0	2/9
2	2/9	0	7/9	0
3	0	2/3	0	1/3

To see that the zeros will persist, note that if initially we have an odd number of balls in the left urn, then no matter whether we add or subtract one ball the result will be an even number. Thus, X_n alternates between being odd and even. To see why this prevents convergence note that $p^{2n}(i, i) > 0$, while $p^{2n+1}(i, i) = 0$.

A second thing that can prevent convergence is shown by

Example 4.16 **Reducible chain**

	0	**1**	**2**	**3**
0	2/3	1/3	0	0
1	1/5	4/5	0	0
2	0	0	1/2	1/2
3	0	0	1/6	5/6

In this case if we start at 0 or 1 it is impossible to get to states 2 or 3 and vice versa, so the 2×2 blocks of 0's will persist forever in the matrix.

A remarkable fact about Markov chains (on finite state spaces) is that if we avoid these two problems then there is a unique stationary distribution π and $p^n(i, j) \to \pi(j)$. The two conditions that rule out these problems are

- *p is said to be* **irreducible** *if for each i and j it is possible to get from i to j, that is, $p^m(i, j) > 0$ for some $m \geq 1$.*
- *a state i is said to be* **aperiodic** *if the greatest common divisor of $J_i = \{n \geq 1$ that have $p^n(i, i) > 0\}$ is 1.*

In general, the greatest common divisor of J_i is called the **period** of state i. In the Ehrenfest chain it is only possible to go from i to i in an even number of steps, so all states have period 2. The next example explains why the definition is formulated in terms of the greatest common divisor.

Example 4.17 **Triangle and square.** The state space is $\{-2, -1, 0, 1, 2, 3\}$ and the transition probability is

	−2	**−1**	**0**	**1**	**2**	**3**
−2	0	0	1	0	0	0
−1	1	0	0	0	0	0
0	0	1/2	0	1/2	0	0
1	0	0	0	0	1	0
2	0	0	0	0	0	1
3	0	0	1	0	0	0

In words, from 0 we are equally likely to go to 1 or -1. From -1 we go with probability 1 to -2 and then back to 0, from 1 we go to 2, then to 3, and back to 0. The name refers to the fact that $0 \to -1 \to -2 \to 0$ is a triangle and $0 \to 1 \to 2 \to 3 \to 0$ is a square.

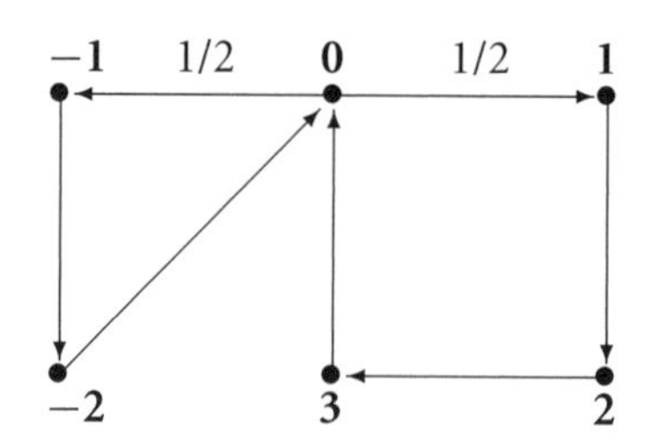

If we note that (i) $3, 4 \in J_0$, (ii) J_0 is closed under addition, and (iii) once J_0 contains three consecutive integers, it will contain the rest then we see

$$J_0 = \{3, 4, 6, 7, 8, 9, \dots\}$$

so the greatest common divisor of J_0 is 1. In this case and in general for aperiodic states J_0 contains all integers beyond some point.

With the key definitions made we can now state the

Convergence theorem. *If p is irreducible and has an aperiodic state then there is a unique stationary distribution π and for any i and j*

$$p^n(i, j) \to \pi(j) \quad \text{as } n \to \infty \tag{4.10}$$

An easy, but important, special case is

Corollary. *If for some n, $p^n(i, j) > 0$ for all i and j, then there is a unique stationary distribution π and*

$$p^n(i, j) \to \pi(j) \quad \textit{as } n \to \infty \tag{4.11}$$

Proof. In this case p is irreducible since it is possible to get from any state to any other in n steps. All states are aperiodic since we also have $p^{n+1}(i, j) > 0$, so $n, n+1 \in J_i$ and hence the greatest common divisor of J_i is 1. □

The corollary with $n = 1$ shows that the convergence theorem applies to the Wright–Fisher model with mutation, weather chain, social mobility, and brand preference chains that are Examples 4.2, 4.5, 4.6, and 4.7. The convergence theorem does not apply to the gambler's ruin chain (Example 4.1) or the Wright–Fisher model with no mutations since they have absorbing states and hence are not irreducible. We have already noted that the Ehrenfest chain (Example 4.3) does not converge since all states have period 2. This leaves the inventory chain (Example 4.4):

	0	**1**	**2**	**3**	**4**	**5**
0	0	0	0.1	0.2	0.4	0.3
1	0	0	0.1	0.2	0.4	0.3
2	0.3	0.4	0.3	0	0	0
3	0.1	0.2	0.4	0.3	0	0
4	0	0.1	0.2	0.4	0.3	0
5	0	0	0.1	0.2	0.4	0.3

We have two results we can use.

Checking (4.10). To check irreducibility we note that starting from 0, 1, or 5, we can get to 2, 3, 4, and 5 in one step and to 0 and 1 in two steps by going through 2 or 3. From 2 or 3 we can get to 0, 1, and 2 in one step and to 3, 4, and 5 in two steps by going through 0. Finally from 4 we can get to 1, 2, 3, and 4 in one step and to 0 or 5 in two steps by going through 2 or 1 respectively. To check aperiodicity, we note that $p(5, 5) > 0$, so 5 is aperiodic.

Checking (4.11). With a calculator we can compute p^2:

	0	**1**	**2**	**3**	**4**	**5**
0	0.05	0.12	0.22	0.28	0.24	0.09
1	0.05	0.12	0.22	0.28	0.24	0.09
2	0.09	0.12	0.16	0.14	0.28	0.21
3	0.15	0.22	0.27	0.15	0.12	0.09
4	0.10	0.19	0.29	0.26	0.13	0.03
5	0.05	0.12	0.22	0.28	0.24	0.09

All entries are positive, so (4.11) applies.

Doubly stochastic chains. A Markov chain is defined by the condition that $\sum_j p(i, j) = 1$. Suppose that in addition, we have $\sum_i p(i, j) = 1$. If the chain has N states then the stationary distribution is $\pi(i) = 1/N$ since

$$\sum_i \pi(i)p(i, j) = \frac{1}{N} \sum_i p(i, j) = \frac{1}{N}$$

To illustrate this consider

Example 4.18

Tiny board game. Consider a circular board game with only six spaces $\{0, 1, 2, 3, 4, 5\}$. On each turn we decide how far to move by flipping two coins and then moving one space for each heads. Here, we consider 5 to be adjacent to 0, so if we are there and get two heads then the result is $5 + 2 \bmod 6 = 1$, where $i + k \bmod 6$ is the remainder when $i + k$ is divided by 6. In this case the transition probability is

	0	1	2	3	4	5
0	1/4	1/2	1/4	0	0	0
1	0	1/4	1/2	1/4	0	0
2	0	0	1/4	1/2	1/4	0
3	0	0	0	1/4	1/2	1/4
4	1/4	0	0	0	1/4	1/2
5	1/2	1/4	0	0	0	1/4

It is clear that the columns add to one, so the stationary distribution is uniform. To check the hypothesis of the convergence theorem, we note that after 3 turns we will have moved between 0 and 6 spaces, so $p^3(i, j) > 0$.

Example 4.19

Mathematician's monopoly. The game monopoly is played on a game board that has 40 spaces arranged around the outside of a square. The squares have names such as *Reading Railroad* and *Park Place*, but we will number the squares 0 (*Go*), 1 (*Baltic Avenue*), ..., 39 (*Boardwalk*). In Monopoly you roll two dice and move forward a number of spaces equal to the sum. For the moment, we will ignore things such as *Go to Jail*, *Chance*, and other squares that make the transitions complicated and formulate the dynamics as following. Let r_k be the probability that the sum of two dice is k ($r_2 = 1/36$, $r_3 = 2/36, \ldots, r_7 = 6/36, \ldots$, $r_{12} = 1/36$) and let

$$p(i, j) = r_k \quad \text{if } j = i + k \bmod 40$$

where $i + k \bmod 40$ is the remainder when $i + k$ is divided by 40. To explain suppose that we are sitting on *Park Place* $i = 37$ and roll $k = 6$. $37 + 6 = 43$ but when we divide by 40 the remainder is 3, so $p(37, 3) = r_6 = 5/36$.

This example is larger but has the same structure as the previous example. Each row has the same entries but shifts 1 unit to the right each time with the number that goes off the right edge emerging in the 0 column. This structure implies that each entry in the row appears once in each column and hence the sum of the entries in the column is 1, and the stationary distribution is uniform. To check the hypothesis of the convergence theorem note that in four rolls you can move forward by 8–48 squares, so $p^4(i, j) > 0$ for all i and j.

Example 4.20

Real Monopoly has two complications:

1. Square 30 is *Go to Jail*, which sends you to square 10. You can buy your way out of jail but in the results we report next, we assume that you are cheap. If you roll a double then you get out for free. If you don't get doubles in three tries you have to pay.

2. There are three *Chance* squares at 7, 12, and 36 (diamonds on the graph), and three *Community Chest* squares at 2, 17, 33 (squares on the graph), where you draw a card, which can send you to another square.

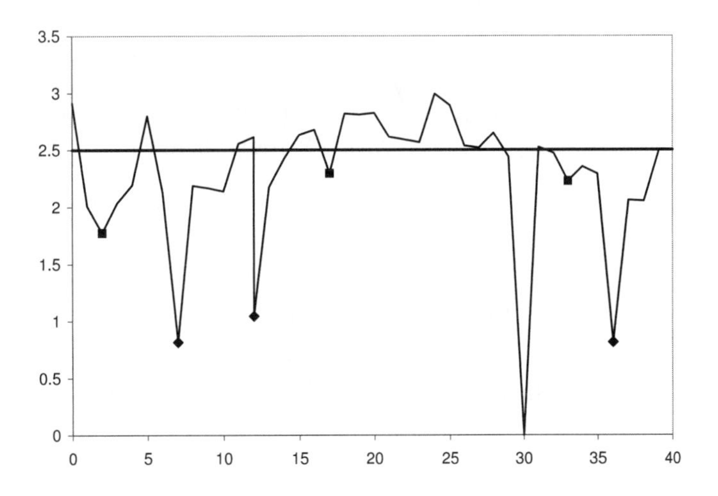

The graph gives the long-run frequencies of being in different squares on the Monopoly board at the end of your turn, as computed by simulation. To make things easier to see we have removed the 9.46% chance of being *In Jail* to make the probabilities easier to see. The value reported for 10 is the 2.14% probability of *Just Visiting Jail*, that is, being brought there by the roll of the dice. Square 30, *Go to Jail*, has probability 0 for the obvious reasons. The other three lowest values occur for *Chance* squares. Due to the transition from 30 to 10, frequencies for squares near 20 are increased relative to the average of 2.5%, while those after 30 or before 10 are decreased. Squares 0 (*Go*) and 5 (*Reading Railroad*) are exceptions to this trend since there are chance cards that instruct you to go there.

In Monopoly when you land on Go you get an extra \$200. This motivates the following.

Example 4.21

Landing on Go. What is the long-run probability that *during one trip around the board* in mathematician's Monopoly that we land on Go? We have italicized a phrase in the last question to emphasize that this is different from the question answered in Example 4.19: the long-run fraction of time we spend at 0 is $1/40$.

Intuitive solution. Consider instead an infinite board with squares $0, 1, 2, \ldots,$ and make the connection with the circular board by thinking of Go as $0, 40, 80, \ldots$. Rolling two dice we move an average of 7 squares per turn, so in the long run we visit $1/7$ of the squares.

Formal solution. Let $X_n = 0$ if we visit square n on the infinite board. For the squares m we do not visit, we define X_m to be the number of steps to the next square we hit. For example if the first three rolls were 4, 6, and 3.

X_n	0	3	2	1	0	5	4	3	2	1	0	2	1	0
n	0	1	2	3	4	5	6	7	8	9	10	11	12	13

Let q_k be the probability we move forward k squares on one roll and suppose $q_0 = 0$. Then X_n is a Markov chain with

$$p_{j,j-1} = 1 \quad j > 0$$

$$p_{0,k-1} = q_k$$

To check the second note that in the example the first roll is 4 but $X_1 = 3$. Let $r_k = \sum_{j=k+1}^{\infty} q_j$. We claim that $rp = r$. To check this, note that to end at $k \geq 0$ there are two possibilities: come down from $k+1$ or jump up from 0, so

$$(rp)_k = r_{k+1} \cdot 1 + r_0 \cdot q_{k+1} = r_k$$

since $r_0 = \sum_{j=1}^{\infty} q_j = 1$ and $q_{k+1} + \sum_{j=k+2}^{\infty} q_j = r_k$.

To turn r into a stationary distribution we need to divide by $\sum_{k=0}^{\infty} r_k$. To evaluate the sum we write

$$\sum_{k=0}^{\infty} \sum_{j=k+1}^{\infty} q_j = \sum_{j=0}^{\infty} \sum_{k=0}^{j-1} q_j = \sum_{j=0}^{\infty} j q_j = \mu$$

Thus the stationary distribution is $\pi_k = r_k / \mu$.

When q_k is the probability that the sum of two dice is k, then X_n is irreducible on the state space $\{0, 1, \ldots, 11\}$. $p^k(0,0) = q_k > 0$ for $k = 2, 3$, so X_n is a periodic and we can conclude $P(X_n = 0) \to 1/\mu = 1/7$.

4.5 Gambler's ruin

As we have said earlier, the long-run behavior of the gambler's ruin chain and the Wright–Fisher model with no mutation are not very exciting. After a while the chain enters one of the absorbing states (0 and N) and stays there forever. Our first question is, "What is the probability that the chain gets absorbed at N before hitting 0?," that is, "What is the probability that the gambler avoids ruin?"

Example 4.22

Flipping coins. A gambler is betting \$1 each time on the outcome of the flip of a fair coin. He has \$10 and will stop when he has \$25. What is the probability that he will reach his goal before he runs out of money?

Even though we are interested only in what happens when we start at 10, to solve this problem we must compute $h(x) =$ the probability of reaching N before 0 starting from x for all $0 < x < N$. Of course, $h(0) = 0$ and $h(N) = 1$. For $0 < x < N$, considering what happens on the first step gives

$$h(x) = \frac{1}{2}h(x-1) + \frac{1}{2}h(x+1)$$

In words, $h(x)$ is the average of $h(x-1)$ and $h(x+1)$. Multiplying by 2, moving $h(x-1)$ to the left and one of the $h(x)$'s to the right, we have

$$h(x) - h(x-1) = h(x+1) - h(x)$$

This says that the slope of h is constant or the graph of h is a straight line. Since $h(0) = 0$ and $h(N) = 1$, the slope must be $1/N$ and

$$h(x) = \frac{x}{N} \tag{4.12}$$

Thus the answer to our question is 10/25.

To see that this is reasonable, note that since we are playing a fair game (that is, the average winnings on any play is 0) the average amount of money we have at any time is the \$10 we started with. When the game ends we will have \$25 with probability p and \$0 with probability $1 - p$. For the expected value to be \$10, we must have $p = 10/25$. This calculation extends easily to the general case: when the game ends we will have $\$N$ with probability p and \$0 with probability $1 - p$. If we start with $\$x$, then the expected value at the end should also be $\$x$ and we must have $p = x/N$.

Example 4.23

Wright–Fisher model. As described in Example 4.2, if we let X_n be the number of A alleles at time n, then X_n is a Markov chain with transition probability

$$p(x, y) = \binom{N}{y}\left(\frac{x}{N}\right)^y\left(1 - \frac{x}{N}\right)^{N-y}$$

0 and N are absorbing states. What is the probability that the chain ends up in N starting from x.

Extending the reasoning in the previous example, we see that if $h(x)$ is the probability of getting absorbed in state N when we start in state x, then $h(0) = 0$, $h(N) = 1$, and for $0 < x < N$,

$$h(x) = \sum_y p(x, y)h(y) \tag{4.13}$$

In words if we jump from x to y on the first step then our absorption probability becomes $h(y)$.

In the Wright–Fisher model, $p(x, y)$ is the binomial distribution for N independent trials with success probability x/N, so the expected number of As after the transition is x; that is, the expected number of As remains constant in time. Using the reasoning from the previous example, we guess

$$h(y) = \frac{y}{N}$$

Clearly, $h(0) = 0$ and $h(N) = 1$. To check (4.13) we note that

$$\sum_y p(x, y)\frac{y}{N} = \frac{x}{N}$$

since the mean of the binomial is x.

The formula $h(y) = y/N$ says that if we start with y A's in the population then the probability that we will end with a population of all A's (an event called "fixation" in genetics) is y/N, the fraction of the population that is A. The case $y = 1$ is a famous result due to Kimura: the probability of fixation of a new mutation is $1/N$. If we suppose that each individual experiences mutations at rate μ, then since there are N individuals, new mutations occur at a total rate $N\mu$. Since each mutation achieves fixation with probability $1/N$, the rate at which mutations become fixed is μ independent of the size of population.

Example 4.24

Roulette. Suppose now that the gambler is playing roulette where he will win \$1 with probability $p = 18/38$ and lose \$1 with probability $1 - p = 20/38$ each time. He has \$25 and will stop when he has \$50. What is the probability that he will reach his goal before he runs out of money? Note if this was a fair game his success probability would be 0.5.

Again we let $h(x) =$ the probability of reaching N before 0 starting from x. Of course $h(0) = 0$ and $h(N) = 1$. For $0 < x < N$, (4.13) implies

$$h(x) = (1 - p)h(x - 1) + ph(x + 1)$$

Moving $(1 - p)h(x - 1)$ to the left and $ph(x)$ to the right, we have

$$(1 - p)(h(x) - h(x - 1)) = p(h(x + 1) - h(x))$$

which rearranges to

$$h(x + 1) - h(x) = \frac{1 - p}{p}(h(x) - h(x - 1)) \qquad (\star)$$

We know that $h(0) = 0$. We don't know $h(1)$, so we let $h(1) = c$. Using ($\star$) repeatedly we have

$$h(2) - h(1) = \frac{1-p}{p}(h(1) - h(0)) = \left(\frac{1-p}{p}\right) c$$

$$h(3) - h(2) = \frac{1-p}{p}(h(2) - h(1)) = \left(\frac{1-p}{p}\right)^2 c$$

$$h(4) - h(3) = \frac{1-p}{p}(h(3) - h(2)) = \left(\frac{1-p}{p}\right)^3 c$$

From this it should be clear that

$$h(x+1) - h(x) = \left(\frac{1-p}{p}\right)^x c$$

Writing $r = (1-p)/p$ to simplify and recalling $h(0) = 0$, we have

$$h(y) = h(y) - h(0) = \sum_{x=1}^{y} h(x) - h(x-1)$$

$$= (r^{y-1} + r^{y-2} + \cdots + r + 1)c = \frac{r^y - 1}{r - 1} \cdot c$$

since $(r^{y-1} + r^{y-2} + \cdots + r + 1)(r - 1) = r^y - 1$. We want $h(N) = 1$, so we must have $c = (r-1)/(r^N - 1)$. It follows that

$$h(y) = \frac{r^y - 1}{r^N - 1} = \frac{\left(\frac{1-p}{p}\right)^y - 1}{\left(\frac{1-p}{p}\right)^N - 1} \tag{4.14}$$

To see what this says for our roulette example we take $p = 18/38$, $x = 25$, and $N = 50$. In this case, $(1-p)/p = 10/9$, so the probability that we succeed is

$$\frac{(10/9)^{25} - 1}{(10/9)^{50} - 1} = \frac{12.929}{193.03} = 0.067$$

compared to 0.5 for the fair game.

Now let's turn things around and look at the game from the viewpoint of the casino, that is, $p = 20/38$. Suppose that the casino starts with the rather modest capital of $x = 100$. (4.14) implies that the probability that they will reach N before going bankrupt is

$$\frac{(9/10)^{100} - 1}{(9/10)^N - 1}$$

If we let $N \to \infty$, $(9/10)^N \to 0$, so the answer converges to

$$1 - (9/10)^{100} = 1 - 2.656 \times 10^{-5}$$

If we increase the capital to \$200 then the failure probability is squared, since to become bankrupt we must first lose \$100 and then lose our second \$100. In this case the failure probability is incredibly small: $(2.656 \times 10^{-5})^2 = 7.055 \times 10^{-10}$.

4.6 Absorbing chains

In this section we consider general Markov chains with absorbing states. The two questions of interest are "Where does the chain get absorbed?" and "How long does it take to get there?" We begin with a simple example.

Example 4.25

Two-year college. At a local two-year college, 60% of freshmen become sophomores, 25% remain freshmen, and 15% drop out. 70% of sophomores graduate transfer to a four-year college, 20% remain sophomores, and 10% drop out. What fraction of new students eventually graduate?

We use a Markov chain with state space 1 = freshman, 2 = sophomore, G = graduate, and D = dropout. The transition probability is

	1	**2**	**G**	**D**
1	0.25	0.6	0	0.15
2	0	0.2	0.7	0.1
G	0	0	1	0
D	0	0	0	1

Let $h(x)$ be the probability that a student currently in state x eventually graduates. By considering what happens on one step,

$$h(1) = 0.25h(1) + 0.6h(2)$$
$$h(2) = 0.2h(2) + 0.7$$

so $h(2) = 0.7/0.8 = 0.875$ and $h(1) = (0.6)/(0.75)h(2) = 0.7$.

To check these answers we will look at powers of the transition probability

$p^2 =$

	1	**2**	**G**	**D**
1	0.0625	0.27	0.42	0.2475
2	0	0.04	0.84	0.12
G	0	0	1	0
D	0	0	0	1

From this we see that the fraction of freshmen who graduate in 2 years is $p^2(1, G) = 0.42 = 0.6(0.7) = p(1, 2)p(2, G)$. After 3 years

$p^3 =$	1	2	G	D
1	0.015625	0.0915	0.609	0.283875
2	0	0.008	0.868	0.124
G	0	0	1	0
D	0	0	0	1

while after 6 years very few people are still trying to get their degrees

$p^6 =$	1	2	G	D
1	0.000244	0.002161	0.697937	0.283875
2	0	0.000064	0.874944	0.124492
G	0	0	1	0
D	0	0	0	1

Note that $p^6(1, D) \approx 0.7$ and $p^6(2, D) \approx 0.875$.

Example 4.26 **Tennis.** In tennis the winner of a game is the first player to win 4 points, unless the score is 4–3, in which case the game must continue until one player wins by 2 points. Suppose that the game has reached the point where one player is trying to get 2 points ahead to win and that the server will independently win the point with probability 0.6. What is the probability that the server will win the game if the score is tied 3–3? if she is ahead by 1 point? Behind by 1 point?

We formulate the game as a Markov chain in which the state is the difference of the scores. The state space is 2, 1, 0, −1, −2 with 2 (win for server) and −2 (win for opponent). The transition probability is

$p =$	2	1	0	−1	−2
2	1	0	0	0	0
1	0.6	0	0.4	0	0
0	0	0.6	0	0.4	0
−1	0	0	0.6	0	0.4
−2	0	0	0	0	1

If we let $h(x)$ be the probability of the server winning when the score is x, then

$$h(x) = \sum_y p(x, y)h(y)$$

with $h(2) = 1$ and $h(-2) = 0$. This involves solving three equations in three unknowns. The computations become much simpler if we look at

$$p^2 = \begin{array}{c|ccccc} & \mathbf{2} & \mathbf{1} & \mathbf{0} & \mathbf{-1} & \mathbf{-2} \\ \mathbf{2} & 1 & 0 & 0 & 0 & 0 \\ \mathbf{1} & 0.6 & 0.24 & 0 & 0.16 & 0 \\ \mathbf{0} & 0.36 & 0 & 0.48 & 0 & 0.16 \\ \mathbf{-1} & 0 & 0.36 & 0 & 0.24 & 0.4 \\ \mathbf{-2} & 0 & 0 & 0 & 0 & 1 \end{array}$$

From p^2 we see that

$$h(0) = 0.36 + 0.48h(0)$$

so $h(0) = 0.36/0.52 = 0.6923$. By considering the outcome of the first point we see that $h(1) = 0.6 + 0.4h(0) = 0.8769$ and $h(-1) = 0.6h(0) = 0.4154$.

General solution. Suppose that the server wins each point with probability w. If the game is tied then after 2 points, the server will have won with probability w^2, lost with probability $(1 - w)^2$, and returned to a tied game with probability $2w(1 - w)$, so $h(0) = w^2 + 2w(1 - w)h(0)$. Since $1 - 2w(1 - w) = w^2 + (1 - w)^2$, solving gives

$$h(0) = \frac{w^2}{w^2 + (1 - w)^2}$$

The next graph gives the probability of winning a tied game as a function of the probability of winning a point.

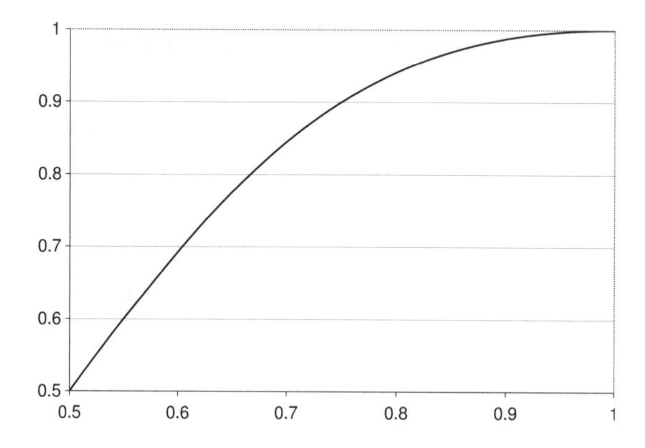

4.6.1 Absorption times

Example 4.27 An office computer is in one of three states, working (W), being repaired (R), or scrapped (S). If the computer is working 1 day the probability it will be working the next day is 0.995 and the probability it will need repair is 0.005. If it is being repaired then probability it is working the next day is 0.9, the probability it still needs repair the next day is 0.05, and the probability it will be scrapped is 0.05. What is the average number of working days until a computer is scrapped?

Leaving out the absorbing state of being scrapped the reduced transition probability is

$$\begin{array}{lcc} & \mathbf{W} & \mathbf{R} \\ r = \mathbf{W} & 0.995 & 0.005 \\ \mathbf{R} & 0.90 & 0.05 \end{array}$$

To see that this is enough of the matrix to compute what we want, note that $r^n(W, W)$ gives the probability a computer is working on day n, so $\sum_{n=0}^{\infty} r^n(W, W)$ gives the expected number of days that it is working. To see this, let $Y_n = 1$ if the computer is working on day n and 0 otherwise; then $Y = \sum_{n=0}^{\infty} Y_n$ is the number of days the computer is working

$$EY = E \sum_{n=0}^{\infty} Y_n = \sum_{n=0}^{\infty} EY_n = \sum_{n=0}^{\infty} r^n(W, W)$$

By analogy with the geometric series $\sum_{n=0}^{\infty} x^n = 1/(1-x)$ we can guess that

$$\sum_{n=0}^{\infty} r^n = (I - r)^{-1} \tag{4.15}$$

where $I =$ the identity matrix and $r^0 = I$. To check this we note that

$$(I - r) \sum_{n=0}^{\infty} r^n = \sum_{n=0}^{\infty} r^n - \sum_{n=1}^{\infty} r^n = r^0 = I$$

Computing the inverse

$$\begin{array}{lcc} & \mathbf{W} & \mathbf{R} \\ (I - r)^{-1} = \mathbf{W} & 3{,}800 & 20 \\ \mathbf{R} & 3{,}600 & 20 \end{array}$$

we see that on the average a working computer will work for 3,800 days and will spend 20 days being repaired.

$(I - r)^{-1}$ is sometimes called the **fundamental matrix** because it is the key to computing many quantities for absorbing Markov chains.

Example 4.28

A local cable company classifies its customers according to how many months overdue their bill is: 0, 1, 2, 3. Accounts that are 3 months overdue are discontinued (D) if they are not paid. The company estimates that transitions occur according to the following probabilities:

	0	1	2	3	D
0	0.9	0.1	0	0	0
1	0.8	0	0.2	0	0
2	0.7	0	0	0.3	0
3	0.6	0	0	0	0.4
D	0	0	0	0	1

What is the expected number of months for a new customer (that is, one who starts in state 0) to have their service discontinued?

Let r be the reduced 4×4 matrix of transitions between the nonabsorbing states, 0–3.

$$(I - r)^{-1} = \begin{matrix} 416.66 & 41.66 & 8.33 & 2.5 \\ 406.66 & 41.66 & 8.33 & 2.5 \\ 366.66 & 36.66 & 8.33 & 2.5 \\ 250 & 25 & 5 & 2.5 \end{matrix}$$

The first row gives the expected number of visits to each of the four states starting from 0, so the expected time is $416.66 + 41.66 + 8.33 + 2.5 = 469.16$.

Returning to our first two examples:

Example 4.29

Two-year college. Consider the transition probability in Example 4.25. How many years on the average does it take for a freshman to graduate or drop out?

Removing the absorbing states from the transition probability, we obtain the reduced matrix

$$\begin{matrix} & \mathbf{1} & \mathbf{2} \\ r = \mathbf{1} & 0.25 & 0.6 \\ \mathbf{2} & 0 & 0.2 \end{matrix}$$

From this we compute

$$\begin{matrix} & \mathbf{1} & \mathbf{2} \\ (I - r)^{-1} = \mathbf{1} & 1.33 & 1 \\ \mathbf{2} & 0 & 1.25 \end{matrix}$$

so, on the average, a freshman takes $1.33 + 1 = 2.33$ years to either graduate or drop out.

Example 4.30

Tennis. Consider the transition probability in Example 4.26. Suppose that game is tied 3–3? How many more points do we expect to see before the game ends?

Removing the absorbing states from the transition probability, we obtain the reduced matrix

$r =$	**1**	**0**	**−1**
1	0	0.4	0
0	0.6	0	0.4
−1	0	0.6	0

From this we compute

$(I-r)^{-1} =$	**1**	**0**	**−1**
1	1.4615	0.7692	0.307
0	1.1538	1.09230	0.769
−1	0.6923	1.1538	1.461

so, on the average, a tied game requires $1.1538 + 1.09230 + 0.769 = 3.8458$ points to be completed.

The fundamental matrix can also be used to compute the probability of winning the game. To see this we note that in order to end in state 2, the chain must wander among the nonabsorbing states for some number of times n and then jump from some state y to state 2; that is,

$$h(x) = \sum_y \sum_{n=0}^{\infty} r^n(x,y)p(y,2) = (I-r)^{-1}(x,y)p(y,2)$$

In the case of the tennis chain, $p(y,2) = 0$ unless $y = 1$, and in this case $p(1,2) = 0.6$, so

$$h(x) = 0.6(I-p)^{-1}(x,1)$$

Multiplying the first column of the previous matrix by 0.6 we get the answers we found in Example 4.26:

$$0.8769 \qquad 0.6923 \qquad 0.4154$$

Example 4.31 **Waiting for coin patterns.** If we flip a coin repeatedly, what is the average number of tosses until we see HH? HT?

If we let (X_{n-1}, X_n) be the outcomes of the last two tosses then this is a Markov chain with transition probability

	HH	**HT**	**TH**	**TT**
HH	1/2	1/2	0	0
HT	0	0	1/2	1/2
TH	1/2	1/2	0	0
TT	0	0	1/2	1/2

To compute the expected time to reach HH, T_{HH}, we make a reduced matrix r_1 by eliminating its row and column and then subtract from the identity to get

$$I - r_1 = \begin{array}{cccc} & \textbf{HT} & \textbf{TH} & \textbf{TT} \\ \textbf{HT} & 1 & -1/2 & -1/2 \\ \textbf{TH} & -1/2 & 1 & 0 \\ \textbf{TT} & 0 & -1/2 & 1/2 \end{array}$$

From this we compute

$$(I - r_1)^{-1} = \begin{array}{cccc} & \textbf{HT} & \textbf{TH} & \textbf{TT} \\ \textbf{HT} & 2 & 2 & 2 \\ \textbf{TH} & 1 & 2 & 1 \\ \textbf{TT} & 1 & 2 & 3 \end{array}$$

From this we see $E_{\text{HT}} T_{\text{HH}} = 6$, $E_{\text{TH}} T_{\text{HH}} = 4$, and $E_{\text{TT}} T_{\text{HH}} = 6$. It takes two tosses to get the chain started. If we get HH on the first two we are done. In the other three cases we have the waiting time computed earlier, so

$$E\,T_{\text{HH}} = 2 + \frac{1}{4}(6 + 4 + 6) = 6$$

We can check this by noting that if we start with T or HT, then we have made no progress, but if the first two tosses are HH we are done. From this we see that

$$E\,T_{\text{HH}} = \frac{1}{2}(1 + E\,T_{\text{HH}}) + \frac{1}{4}(2 + E\,T_{\text{HH}}) = \frac{1}{4} \cdot 2$$

so $(1/4)E\,T_{\text{HH}} = 1.5$ and again we find $E\,T_{\text{HH}} = 6$.

It is clear from symmetry that $E\,T_{\text{TT}} = 6$, so the only remaining case is $E\,T_{\text{HT}}$. To compute the expected time to reach HT, we make a reduced matrix r_2 by eliminating its row and column and then subtract from the identity to get

$$I - r_2 = \begin{array}{cccc} & \textbf{HH} & \textbf{TH} & \textbf{TT} \\ \textbf{HH} & 1/2 & 0 & 0 \\ \textbf{TH} & -1/2 & 1 & 0 \\ \textbf{TT} & 0 & -1/2 & 1/2 \end{array}$$

From this we compute

$$(I - r_2)^{-1} = \begin{array}{cccc} & \textbf{HH} & \textbf{TH} & \textbf{TT} \\ \textbf{HH} & 2 & 0 & 0 \\ \textbf{TH} & 1 & 1 & 0 \\ \textbf{TT} & 1 & 1 & 2 \end{array}$$

From this we see $E_{\text{HH}} T_{\text{HT}} = 2$, $E_{\text{TH}} T_{\text{HT}} = 2$, and $E_{\text{TT}} T_{\text{HT}} = 4$, and

$$E\,T_{\text{HT}} = 2 + \frac{1}{4}(2 + 2 + 4) = 4$$

This is the answer we should have expected to hold in general: along the sequence 1/4 of the pairs are HT, so the average distance between them is 4. We get a different answer for $E\,T_{\text{HH}}$ because once one HH occurs then there is probability 1/2 of having another HH one unit of time later. $E\,T_{\text{HH}}$ counts the distance between nonoverlapping occurrences, so

$$4 = (1/2) \cdot 1 + (1/2) \cdot E\,T_{\text{HH}}$$

Solving we have a third proof that $E\,T_{\text{HH}} = 6$.

Example 4.32 **Waiting for coin patterns, II.** We now consider patterns of length 3. By symmetry we only need to consider the four that start with H. Using the reasoning from the end of the previous example,

$$8 = (1/2) \cdot 1 + (1/2) \cdot E\,T_{\text{HHH}}$$
$$8 = (1/4) \cdot 2 + (3/4) \cdot E\,T_{\text{HTH}}$$

and solving gives

$$E\,T_{\text{HHH}} = 14 \qquad E\,T_{\text{HTH}} = 10$$

In the other cases, the three tosses are not useful for getting another occurrence so $E\,T_{\text{HHT}} = E\,T_{\text{HTT}} = 8$.

To check the result for $E\,T_{\text{HHH}}$, we note that if we start with T, HT, or HHT, then we have made no progress, but if the first three tosses are HHH we are done. From this we see that

$$E\,T_{\text{HHH}} = \frac{1}{2}(1 + E\,T_{\text{HHH}}) + \frac{1}{4}(2 + E\,T_{\text{HHH}}) + \frac{1}{8}(3 + E\,T_{\text{HHH}}) + \frac{1}{8} \cdot 3$$

so $(1/8)E\,T_{\text{HHH}} = 1.75$ and again we find $E\,T_{\text{HHH}} = 14$.

To compute $E\,T_{\text{HTH}}$ we begin by writing down the transition probability for the last three flips (X_{n-2}, X_{n-1}, X_n):

	HHH	HHT	HTH	HTT	THH	THT	TTH	TTT
HHH	1/2	1/2	0	0	0	0	0	0
HHT	0	0	1/2	1/2	0	0	0	0
HTH	0	0	0	0	1/2	1/2	0	0
HTT	0	0	0	0	0	0	1/2	1/2
THH	1/2	1/2	0	0	0	0	0	0
THT	0	0	1/2	1/2	0	0	0	0
TTH	0	0	0	0	1/2	1/2	0	0
TTT	0	0	0	0	0	0	1/2	1/2

Removing the row and column for HTH, then subtracting from I the matrix to be inverted is

$$\begin{matrix} 1/2 & -1/2 & 0 & 0 & 0 & 0 & 0 \\ 0 & 1 & -1/2 & 0 & 0 & 0 & 0 \\ 0 & 0 & 1 & 0 & 0 & -1/2 & -1/2 \\ -1/2 & -1/2 & 0 & 1 & 0 & 0 & 0 \\ 0 & 0 & -1/2 & 0 & 1 & 0 & 0 \\ 0 & 0 & 0 & -1/2 & -1/2 & 1 & 0 \\ 0 & 0 & 0 & 0 & 0 & -1/2 & 1/2 \end{matrix}$$

The inverse is

$$\begin{matrix} 2.5 & 1.5 & 1 & 0.5 & 0.5 & 1 & 1 \\ 0.5 & 1.5 & 1 & 0.5 & 0.5 & 1 & 1 \\ 1 & 1 & 2 & 1 & 1 & 2 & 2 \\ 1.5 & 1.5 & 1 & 1.5 & 0.5 & 1 & 1 \\ 0.5 & 0.5 & 1 & 0.5 & 1.5 & 1 & 1 \\ 1 & 1 & 1 & 1 & 1 & 2 & 1 \\ 1 & 1 & 1 & 1 & 1 & 2 & 3 \end{matrix}$$

From this we conclude that the expected waiting times are

$x =$	HHH	HHT	HTT	THH	THT	TTH	TTT
$E_x T_{\text{HTH}}$	8	6	10	8	6	8	10

As in the previous example it takes three tosses to get the chain started and at that point all 8 states are equally likely, so

$$E\,T_{\text{HTH}} = 3 + \frac{1}{8}(8 + 6 + 10 + 8 + 6 + 8 + 10) = 3 + \frac{56}{8} = 10$$

Removing the row and column for HHT, then subtracting from I the matrix to be inverted is

$$\begin{matrix} 1/2 & 0 & 0 & 0 & 0 & 0 & 0 \\ 0 & 1/2 & -1/2 & 0 & 0 & 0 & 0 \\ 0 & 0 & 1 & 0 & 0 & -1/2 & -1/2 \\ -1/2 & 0 & 0 & 1 & 0 & 0 & 0 \\ 0 & -1/2 & -1/2 & 0 & 1 & 0 & 0 \\ 0 & 0 & 0 & -1/2 & -1/2 & 1 & 0 \\ 0 & 0 & 0 & 0 & 0 & -1/2 & 1/2 \end{matrix}$$

which differs from the previous one only in the second row and second column. This small change makes a big difference in the inverse:

$$\begin{matrix} 2 & 0 & 0 & 0 & 0 & 0 & 0 \\ 1 & 1.5 & 0.5 & 1 & 1 & 0.5 & 0.5 \\ 1 & 0.5 & 1.5 & 1 & 1 & 1.5 & 1.5 \\ 1 & 0 & 0 & 1 & 0 & 0 & 0 \\ 1 & 1 & 1 & 1 & 2 & 1 & 1 \\ 1 & 0.5 & 0.5 & 1 & 1 & 1.5 & 0.5 \\ 1 & 0.5 & 0.5 & 1 & 1 & 1.5 & 2.5 \end{matrix}$$

From this we conclude that the expected waiting times are

$x =$	HHH	HHT	HTT	THH	THT	TTH	TTT
$E_x T_{\text{HHT}}$	2	6	8	2	8	6	8

To see that the first and fourth answers are sensible, note that if the initial state is HHH or THH, we will have HHT the first time we get a T. As in the previous example it takes three tosses to get the chain started and at that point all 8 states are equally likely, so

$$E\,T_{\text{HHT}} = 3 + \frac{1}{8}(2 + 6 + 8 + 2 + 8 + 6 + 8) = 3 + \frac{40}{8} = 8$$

We could have saved ourselves a lot of work if we had noted that in order to get HHT we need to first get HH, which has an expected waiting time of 6, and then wait for a T, which has a waiting time of 2, so $E\,T_{\text{HHT}} = 6 + 2$. To treat HTT this way, note that we need to first get HT, which has an expected waiting time of 4. If we get a T, then we are done. If not, we have to wait again for an HT and the average number of times we have to repeat the process is 2, so $E_{\text{HTT}} = 4 \cdot 2$.

4.7 Exercises

Transition probabilities

1. What values of x, y, z will make these matrices transition probabilities:

$$\text{(a)}\ \begin{matrix} 0.5 & 0.1 & x \\ y & 0.2 & 0.4 \\ 0.3 & z & 0.1 \end{matrix} \qquad \text{(b)}\ \begin{matrix} x & 0.1 & 0.7 \\ 0.2 & 0.3 & y \\ 0.6 & z & 0.2 \end{matrix}$$

2. A red urn contains 2 red marbles and 3 blue marbles. A blue urn contains 1 red marble and 4 blue marbles. A marble is selected from an urn, the marble

is returned to the urn from which it was drawn, and the next marble is drawn from the urn with the color that was drawn. (a) Write the transition probability for this chain. (b) Suppose the first marble is drawn from the red urn. What is the probability that the third one will be drawn from the blue urn?

3. At Llenroc College, 63% of freshmen who are premed switch to a liberal arts major, while 18% of liberal arts majors switch to being premed. If the incoming freshman class is 60% premed and 40% liberal arts majors, what fraction graduate as premed?

4. A person is flipping a coin repeatedly. Let X_n be the outcome of the two previous coin flips at time n; for example, the state might be HT to indicate that the last flip was T and the one before that was H. (a) Compute the transition probability for the chain. (b) Find p^2.

5. A taxicab driver moves between the airport A and two hotels B and C according to the following rules. If he is at the airport, he will go to one of the two hotels next with equal probability. If at a hotel then he returns to the airport with probability 3/4 and goes to the other hotel with probability 1/4. (a) Find the transition matrix for the chain. (b) Suppose the driver begins at the airport at time 0. Find the probability for each of his three possible locations at time 2 and the probability that he is at hotel B at time 3.

6. Consider a gambler's ruin chain with $N = 4$. That is, if $1 \le i \le 3$, $p(i, i + 1) = 0.4$, and $p(i, i - 1) = 0.6$, but the endpoints are absorbing states: $p(0, 0) = 1$ and $p(4, 4) = 1$. Compute $p^3(1, 4)$ and $p^3(1, 0)$.

7. An outdoor restaurant in a resort town closes when it rains. From past records it was found that from May to September, when it rains one day the probability that it rains the next is 0.4; when it does not rain one day it rains the next with probability 0.1. (a) Write the transition matrix. (b) If it rained on Thursday what is the probability that it will rain on Saturday? On Sunday?

8. Market research suggests that in a 5-year period 8% of people with cable television will get rid of it and 26% of those without it will sign up for it. Compare the predictions of the Markov chain model with the following data on the fraction of people with cable TV: 56.4% in 1990, 63.4% in 1995, and 68.0% in 2000.

9. A sociology professor postulates that in each decade 8% of women in the workforce leave it and 20% of the women not in it begin to work. Compare the predictions of his model with the following data on the percentage of women working: 43.3% in 1970, 51.5% in 1980, 57.5% in 1990, and 59.8% in 2000.

10. The following transition probability describes the migration patterns of birds between three habitats:

	1	2	3
1	0.75	0.15	0.10
2	0.07	0.85	0.08
3	0.05	0.15	0.80

If there are 1,000 birds in each habitat at the beginning of the first year, how many do we expect to be in each habitat at the end of the year? At the end of the second year?

Convergence to equilibrium

11. A car rental company has rental offices at both Kennedy and LaGuardia airports. Assume that a car rented at one airport must be returned to one of the two airports. If the car was rented at LaGuardia the probability that it will be returned there is 0.8; for Kennedy the probability is 0.7. Suppose that we start with 1/2 of the cars at each airport and that each week all of the cars are rented once. (a) What is the fraction of cars at LaGuardia Airport at the end of the first week? (b) At the end of the second? (c) In the long run?

12. The 1990 census showed that 36% of the households in the District of Columbia were homeowners, while the remainder were renters. During the next decade 6% of the homeowners became renters and 12% of the renters became homeowners. (a) What percentage were homeowners in 2000? In 2010? (b) If these trends continue what will be the long-run fraction of homeowners?

13. Most railroad cars are owned by individual railroad companies. When a car leaves its home railroad's trackage, it becomes part of the national pool of cars and can be used by other railroads. A particular railroad found that each month 15% of its boxcars on its home trackage left to join the national pool and 40% of its cars in the national pool were returned to its home trackage. A company begins on January 1 with all its cars on its home trackage. What fraction will be there on March 1? At the end of the year? In the long run what fraction of a company's cars will be on its home trackage.

14. A rapid transit system has just started operating. In the first month of operation, it was found that 25% of commuters are using the system, while 75% are traveling by automobile. Suppose that each month 10% of transit users go back to using their cars, while 30% of automobile users switch to the transit system. (a) Compute the three-step transition probability p^3. (b) What will be the fractions using rapid transit in the fourth month? (c) In the long run?

15. A regional health study indicates that from one year to the next, 75% of smokers will continue to smoke, while 25% will quit. 8% of those who stopped smoking will resume smoking, while 92% will not. If 70% of the population were smokers in 1995, what fraction will be smokers in 1998? In 2005? In the long run?

16. The town of Mythica has a "free bikes for the people program." You can pick up bikes at the library (L), the coffee shop (C), or the cooperative grocery store (G). The director of the program has determined that if a bike is picked up at the library ends up at the coffee shop with probability 0.2 and at the grocery store with probability 0.3. A bike from the coffee shop will go to the library with probability 0.4 and to the grocery store with probability 0.1. A bike from the grocery store will go to the library or the coffee shop with probability 0.25 each. On Sunday there are an equal number of bikes at each place. (a) What fraction of the bikes are at the three locations on Tuesday? (b) On the next Sunday? (c) In the long run what fraction are at the three locations?

17. Bob eats lunch at the campus food court every week day. He either eats Chinese food, Quesadilla, or Salad. His transition matrix is

	C	Q	S
C	0.15	0.6	0.25
Q	0.4	0.1	0.5
S	0.1	0.3	0.6

He had Chinese food on Monday. What are the probabilities for his three meal choices on Friday (4 days later).

Asymptotic behavior: Two-state chains

18. Census results reveal that in the United States 80% of the daughters of working women work and that 30% of the daughters of nonworking women work. (a) Write the transition probability for this model. (b) In the long run what fraction of women will be working?

19. Three of every four trucks on the road are followed by a car, while only one of every five cars is followed by a truck. What fraction of vehicles on the road are trucks?

20. In a test paper the questions are arranged so that 3/4's of the time a *true* answer is followed by a *true*, while 2/3's of the time a *false* answer is followed by a *false*. You are confronted with a 100-question test paper. Approximately what fraction of the answers will be *true*?

21. When a basketball player makes a shot then he tries a harder shot the next time and hits (H) with probability 0.4 and misses (M) with probability 0.6. When he misses he is more conservative the next time and hits (H) with probability 0.7 and misses (M) with probability 0.3. (a) Write the transition probability for the two-state Markov chain with state space $\{H, M\}$. (b) Find the long-run fraction of time he hits a shot.

22. A regional health study shows that from 1 year to the next 76% of the people who smoked will continue to smoke and 24% will quite. 8% of those who do not smoke will start smoking, while 92% of those who do not smoke will continue to be nonsmokers. In the long run what fraction of people will be smokers?

23. In unprofitable times corporations sometimes suspend dividend payments. Suppose that after a dividend has been paid the next one will be paid with probability 0.9, while after a dividend is suspended the next one will be suspended with probability 0.6. In the long run what is the fraction of dividends that will be paid?

24. A university computer room has 30 terminals. Each day there is a 3% chance that a given terminal will break and a 72% chance that a given broken terminal will be repaired. Assume that the fates of the various terminals are independent. In the long run what is the distribution of the number of terminals that are broken.

Asymptotic behavior: Three or more states

25. A plant species has red, pink, or white flowers according to the genotypes RR, RW, and WW, respectively. If each of these genotypes is crossed with a pink (RW) plant then the offspring fractions are

	RR	RW	WW
RR	0.5	0.5	0
RW	0.25	0.5	0.25
WW	0	0.5	0.5

What is the long-run fraction of plants of the three types?

26. A certain town never has two sunny days in a row. Each day is classified as rainy, cloudy, or sunny. If it is sunny one day then it is equally likely to be cloudy or rainy the next. If it is cloudy or rainy then it remains the same 1/2 of the time, but if it changes it will go to either of the other possibilities with probability 1/4 each. In the long run what proportion of days in this town are sunny? Cloudy? Rainy?

27. A midwestern university has three types of health plans: a health maintenance organization (HMO), a preferred provider organization (PPO), and a traditional fee for service plan (FFS). In 2000, the percentages for the three plans were HMO: 30%, PPO: 25%, and FFS: 45%. Experience dictates that people change plans according to the following transition matrix:

	HMO	PPO	FFS
HMO	0.85	0.1	0.05
PPO	0.2	0.7	0.1
FFS	0.1	0.3	0.6

(a) What will be the percentages for the three plans in 2001? (b) What is the long-run fraction choosing each of the three plans?

28. A sociologist studying living patterns in a certain region determines that the pattern of movement between urban (U), suburban (S), and rural areas (R) is given by the following transition matrix:

	U	S	R
U	0.86	0.08	0.06
S	0.05	0.88	0.07
R	0.03	0.05	0.92

In the long run what fraction of the population will live in the three areas?

29. In a large metropolitan area, commuters either drive alone (A), carpool (C), or take public transportation (T). A study showed that 80% of those who drive alone will continue to do so next year, while 15% will switch to carpooling and 5% will use public transportation. 90% of those who carpool will continue, while 5% will drive alone and 5% will use public transportation. 85% of those who use public transportation will continue, while 10% will carpool and 5% will drive alone. Write the transition probability for the model. In the long run what fraction of commuters will use the three types of transportation?

30. In a particular county voters declare themselves as members of the Republican, Democrat, or Green party. No voters change directly from the Republican to Green party or vice versa. In a given year 15% of Republicans and 5% of Green party members will become Democrats, while 5% of Democrats switch to the Republican party and 10% to the Green party. Write the transition probability for the model. In the long run what fraction of voters will belong to the three parties.

31. (a) Three telephone companies A, B, and C compete for customers. Each year A loses 5% of its customers to B and 20% to C; B loses 15% of its customers

to A and 20% to C; C loses 5% of its customers to A and 10% to B. (a) Write the transition matrix for the model. (b) What is the limiting market share for each of these companies?

32. An auto insurance company classifies its customers in three categories: poor, satisfactory, and preferred. No one moves from poor to preferred or from preferred to poor in 1 year. 40% of the customers in the poor category become satisfactory, 30% of those in the satisfactory category move to preferred, while 10% become poor; 20% of those in the preferred category are downgraded to satisfactory. (a) Write the transition matrix for the model. (b) What is the limiting fraction of drivers in each of these categories?

33. A professor has two lightbulbs in his garage. When both are burned out, they are replaced, and the next day starts with two working lightbulbs. Suppose that when both are working, one of the two will go out with probability 0.02. (Each has probability 0.01 and we ignore the possibility of losing two on the same day.) However, when only one is there, it will burn out with probability 0.05. What is the long-run fraction of time that there is exactly one lightbulb working?

34. An individual has three umbrellas; some at her office and some at home. If she is leaving home in the morning (or leaving work at night) and it is raining, she will take an umbrella, if one is there. Otherwise, she gets wet. Assume that independent of the past, it rains on each trip with probability 0.2. To formulate a Markov chain, let X_n be the number of umbrellas at her current location. (a) Find the transition probability for this Markov chain. (b) Calculate the limiting fraction of time she gets wet.

35. At the end of a month, a large retail store classifies each of its customer's accounts according to current (0), 30–60 days overdue (1), 60–90 days overdue (2), more than 90 days overdue (3). Their experience indicates that the accounts move from state to state according to a Markov chain with transition probability matrix:

	0	**1**	**2**	**3**
0	0.9	0.1	0	0
1	0.8	0	0.2	0
2	0.5	0	0	0.5
3	0.1	0	0	0.9

In the long run what fraction of the accounts are in each category?

36. At the beginning of each day, a piece of equipment is inspected to determine its working condition, which is classified as state 1 = new, 2, 3, or 4 = broken.

We assume the state is a Markov chain with the following transition matrix:

	1	2	3	4
1	0.95	0.05	0	0
2	0	0.9	0.1	0
3	0	0	0.875	0.125

(a) Suppose that a broken machine requires 3 days to fix. To incorporate this into the Markov chain we add states 5 and 6 and suppose that $p(4, 5) = 1$, $p(5, 6) = 1$, and $p(6, 1) = 1$. Find the fraction of time that the machine is working. (b) Suppose now that we have the option of performing preventative maintenance when the machine is in state 3 and that this maintenance takes 1 day and returns the machine to state 1. This changes the transition probability to

	1	2	3
1	0.95	0.05	0
2	0	0.9	0.1
3	1	0	0

Find the fraction of time the machine is working under this new policy.

37. To make a crude model of a forest we might introduce states 0 = grass, 1 = bushes, 2 = small trees, 3 = large trees, and write down a transition matrix such as the following:

	0	1	2	3
0	1/2	1/2	0	0
1	1/24	7/8	1/12	0
2	1/36	0	8/9	1/12
3	1/8	0	0	7/8

The idea behind this matrix is that if left undisturbed a grassy area will see bushes grow, and then small trees, which of course grow into large trees. However, disturbances such as tree falls or fires can reset the system to state 0. Find the limiting fraction of land in each of the states.

38. Five white balls and five black balls are distributed in two urns in such a way that each urn contains five balls. At each step we draw one ball from each urn and exchange them. Let X_n be the number of white balls in the left urn at time n. (a) Compute the transition probability for X_n. (b) Find the stationary distribution and show that it corresponds to picking five balls at random to be in the left urn.

Absorbing chains

39. Two competing companies are trying to buy up all the farms in a certain area to build houses. In each year 10% of farmers sell to company 1, 20% sell to company 2, and 70% keep farming. Neither company ever sells any of the farms that they own. Eventually all the farms will be sold. How many will be owned by company 1?

40. A warehouse has a capacity to hold four items. If the warehouse is neither full nor empty, the number of items in the warehouse changes whenever a new item is produced or an item is sold. Suppose that (no matter when we look) the probability that the next event is "a new item is produced" is 2/3 and that the new event is a "sale" is 1/3. If there is currently one item in the warehouse, what is the probability that the warehouse will become full before it becomes empty.

41. The Macrosoft Company gives each of its employees the title of programmer (P) or project manager (M). In any given year 70% of programmers remain in that position, 20% are promoted to project manager, and 10% are fired (state X). 95% of project managers remain in that position, while 5% are fired. How long on the average does a programmer work before they are fired?

42. At a nationwide travel agency, newly hired employees are classified as beginners (B). Every 6 months the performance of each agent is reviewed. Past records indicate that transitions through the ranks to intermediate (I) and qualified (Q) are according to the following Markov chain, where F indicates workers that were fired:

	B	I	Q	F
B	0.45	0.4	0	0.15
I	0	0.6	0.3	0.1
Q	0	0	1	0
F	0	0	0	1

(a) What fraction are eventually promoted? (b) What is the expected time until a beginner is fired or becomes qualified?

43. At a manufacturing plant, employees are classified as trainee (R), technician (T), or supervisor (S). Writing Q for an employee who quits we model their

progress through the ranks as a Markov chain with transition probability

	R	T	S	Q
R	0.2	0.6	0	0.2
T	0	0.55	0.15	0.3
S	0	0	1	0
Q	0	0	0	1

(a) What fraction of recruits eventually make supervisor? (b) What is the expected time until a trainee quits or becomes supervisor?

44. The two previous problems have the following form:

	1	2	A	B
1	$1-a-b$	a	0	b
2	0	$1-c-d$	c	d
A	0	0	1	0
B	0	0	0	1

Show that (a) the probability of being absorbed in A rather than B is $ac/(a+b)(c+d)$ and (b) the expected time to absorption starting from 1 is $1/(a+b)+a/(a+b)(c+d)$.

45. The Markov chain associated with a manufacturing process may be described as follows: a part to be manufactured will begin the process by entering step 1. After step 1, 20% of the parts must be reworked, that is, returned to step 1, 10% of the parts are thrown away, and 70% proceed to step 2. After step 2, 5% of the parts must be returned to the step 1, 10% to step 2, 5% are scrapped, and 80% emerge to be sold for a profit. (a) Formulate a four-state Markov chain with states 1, 2, 3, and 4, where 3 = a part that was scrapped and 4 = a part that was sold for a profit. (b) Compute the probability that a part is scrapped in the production process.

46. Six children (Dick, Helen, Joni, Mark, Sam, and Tony) play catch. If Dick has the ball he is equally likely to throw it to Helen, Mark, Sam, and Tony. If Helen has the ball she is equally likely to throw it to Dick, Joni, Sam, and Tony. If Sam has the ball he is equally likely to throw it to Dick, Helen, Mark, and Tony. If either Joni or Tony gets the ball, they keep throwing it to each other. If Mark gets the ball he runs away with it. (a) Find the transition probability. (b) Suppose Dick has the ball at the beginning of the game. What is the probability that Mark will end up with it?

5

Continuous Distributions

5.1 Density functions

In many situations random variables can take any value on the real line or in a subset of the real line, such as the nonnegative numbers or the interval $[0, 1]$. For concrete examples, consider the height or weight of a Cornell student chosen at random, or the time it takes a person to drive from Los Angeles to San Francisco. A random variable X is said to have a **continuous distribution** with **density function** f if for all $a \le b$, we have

$$P(a \le X \le b) = \int_a^b f(x)\,dx \tag{5.1}$$

Geometrically, $P(a \le X \le b)$ is the area under the curve f between a and b.

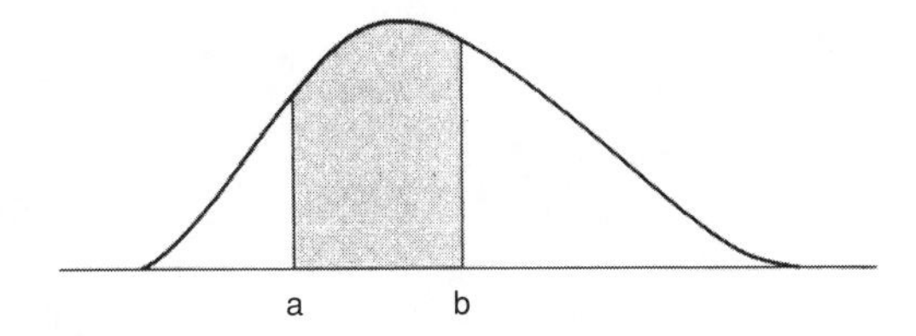

For the purposes of understanding and remembering formulas, it is useful to think of $f(x)$ as $P(X = x)$ even though the last event has probability zero. To explain the last remark and to prove $P(X = x) = 0$, note that taking $a = x$ and $b = x + \Delta x$ in (5.1) we have

$$P(x \le X \le x + \Delta x) = \int_x^{x+\Delta x} f(y)\,dy \approx f(x)\Delta x$$

when Δx is small. Letting $\Delta x \to 0$, we see that $P(X = x) = 0$, but $f(x)$ tells us how likely it is for X to be near x. That is,

$$\frac{P(x \le X \le x + \Delta x)}{\Delta x} \approx f(x)$$

In order for $P(a \le X \le b)$ to be nonnegative for all a and b and for $P(-\infty < X < \infty) = 1$, we must have

$$f(x) \ge 0 \quad \text{and} \quad \int f(x)\, dx = 1 \tag{5.2}$$

Here, and in what follows, if the limits of integration are not specified, the integration is over all values of x from $-\infty$ to ∞. Any function f that satisfies (5.2) is said to be a **density function.**

We are now going to give three important examples of density functions. For simplicity, we will give the values of $f(x)$ where it is positive and omit the phrase "0 otherwise."

Example 5.1

Uniform distribution. Given $a < b$ we define

$$f(x) = \frac{1}{b-a} \qquad a \le x \le b$$

The idea here is that we are picking a value "at random" from (a, b). That is, values outside the interval are impossible, and all those inside have the same probability (density).

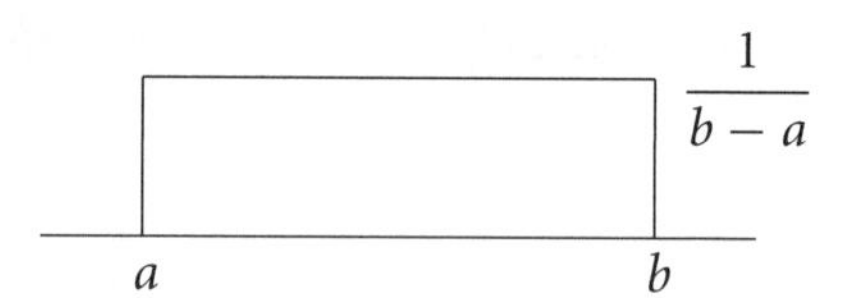

If we set $f(x) = c$ when $a < x < b$ and 0 otherwise then

$$\int f(x)\, dx = \int_a^b c\, dx = c(b-a)$$

So we have to pick $c = 1/(b-a)$ to make the integral 1. The most important special case occurs when $a = 0$ and $b = 1$. Random numbers generated by a computer are typically uniformly distributed on $(0, 1)$. Another case that comes up in applications is $a = -1/2$ and $b = 1/2$. If we take a measurement and round it off to the nearest integer then it is reasonable to assume that the "roundoff error" is uniformly distributed on $(-1/2, 1/2)$.

Example 5.2

Exponential distribution. Given $\lambda > 0$ we define

$$f(x) = \lambda e^{-\lambda x} \qquad x \ge 0$$

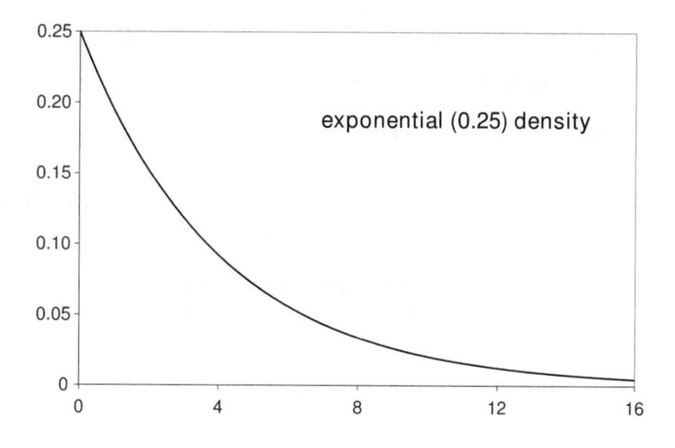

To check that this is a density function, we note that

$$\int_0^\infty \lambda e^{-\lambda x}\, dx = -e^{-\lambda x}\Big|_0^\infty = 0 - (-1) = 1$$

Exponentially distributed random variables often come up as waiting times between events, for example, the arrival times of customers at a bank or ice-cream shop. Sometimes we will indicate that X has an exponential distribution with parameter λ by writing $X = \text{exponential}(\lambda)$.

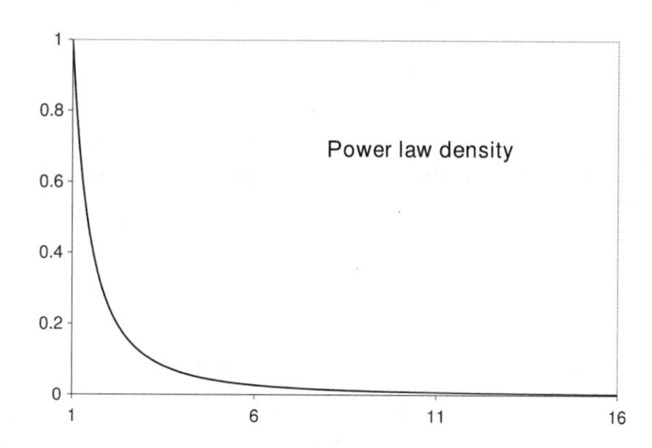

Example 5.3 **Power laws**

$$f(x) = (\rho - 1)x^{-\rho} \quad x \geq 1$$

Here, $\rho > 1$ is a parameter that governs how fast the probabilities go to 0 at ∞.

To check that this is a density function, we note that

$$\int_1^\infty (\rho - 1)x^{-\rho}\, dx = -x^{-(\rho-1)}\Big|_1^\infty = 0 - (-1) = 1$$

These distributions are often used in situations where $P(X > x)$ does not go to 0 very fast as $x \to \infty$. For example, the Italian economist Pareto used them to describe the distribution of family incomes.

A fourth example, and perhaps the most important distribution of all, is the normal distribution. However, because some treatments might skip this chapter, we delay its consideration until Section 6.4.

5.1.1 Expected value

Given a discrete random variable X and a function $r(x)$, the expected value of $r(X)$ is defined by

$$Er(X) = \sum_x r(x)P(X = x).$$

To define the expected value for a continuous random variable, we replace the probability function by the density function and the sum by an integral:

$$Er(X) = \int r(x)\, f(x)\, dx \tag{5.3}$$

As in Section 1.6, if $r(x) = x^k$, EX^k is called the kth moment of X and the variance is defined by

$$\text{var}\,(X) = E(X - EX)^2 = E(X^2) - (EX)^2$$

Example 5.4

Uniform distribution. Suppose X has density function $f(x) = 1/(b - a)$ for $a \le x \le b$. Then

$$EX = \frac{a+b}{2} \qquad \text{var}\,(X) = \frac{(b-a)^2}{12} \tag{5.4}$$

Notice that $(a + b)/2$ is the midpoint of the interval and hence is the natural choice for the average value of X. Since $\text{var}\,(X + c) = \text{var}\,(X)$, the variance depends only on the length of the interval, not its location.

We begin with the case $a = 0$, $b = 1$.

$$EX = \int_0^1 x\, dx = \left.\frac{x^2}{2}\right|_0^1 = 1/2$$

$$E(X^2) = \int_0^1 x^2\, dx = \left.\frac{x^3}{3}\right|_0^1 = 1/3$$

$$\text{var}\,(X) = E(X^2) - (EX)^2 = (1/3) - (1/2)^2 = 1/12$$

To extend to the general case we recall from Section 1.6 that if $Y = c + dX$, then

$$E(Y) = c + d\,EX \quad \text{var}\,(Y) = d^2\,\text{var}\,(X) \tag{5.5}$$

Taking $c = a$ and $d = b - a$,

$$EY = a + \frac{b-a}{2} = \frac{a+b}{2} \quad \text{var}\,(Y) = \frac{(b-a)^2}{12}$$

Example 5.5

Exponential distribution. Suppose X has density function $f(x) = \lambda e^{-\lambda x}$ for $x \geq 0$.

$$EX = 1/\lambda \quad \text{var}\,(X) = 1/\lambda^2 \tag{5.6}$$

To explain the form of the answers, we note that if Y is exponential(1), then $X = Y/\lambda$ is exponential(λ), and then use (5.5) to conclude $EX = EY/\lambda$ and $\text{var}\,(X) = \text{var}\,(Y)/\lambda^2$. Because the mean is inversely proportional to λ, λ is sometimes called the rate.

To compute the moments we need the **integration by parts formula:**

$$\int_a^b g(x)h'(x)\,dx = g(x)h(x)|_a^b - \int_a^b g'(x)h(x)\,dx \tag{5.7}$$

Integrating by parts with $g(x) = x$, $h'(x) = \lambda e^{-\lambda x}$, so $g'(x) = 1$ and $h(x) = -e^{-\lambda x}$.

$$\begin{aligned} EX &= \int_0^\infty x\lambda e^{-\lambda x}\,dx \\ &= -xe^{-\lambda x}|_0^\infty + \int_0^\infty e^{-\lambda x}\,dx = 0 + 1/\lambda \end{aligned}$$

To check the last step note that $-xe^{-\lambda x} = 0$ when $x = 0$ and when $x \to \infty$ $-xe^{-\lambda x} \to 0$ since $e^{-\lambda x} \to 0$ much faster than x grows. The evaluation of the integral follows from the definition of the exponential density, which implies $\int_0^\infty \lambda e^{-\lambda x}\,dx = 1$

To compute $E(X^2)$, we integrate by parts with $g(x) = x^2$ and $h'(x) = \lambda e^{-\lambda x}$, so $g'(x) = 2x$ and $h(x) = -e^{-\lambda x}$.

$$\begin{aligned} E(X^2) = \int_0^\infty x^2\lambda e^{-\lambda x}\,dx &= -x^2 e^{-\lambda x}|_0^\infty + \int_0^\infty 2xe^{-\lambda x}\,dx \\ &= 0 + \frac{2}{\lambda}\int_0^\infty x\lambda e^{-\lambda x}\,dx = \frac{2}{\lambda^2} \end{aligned}$$

by the reasoning in the previous calculation and the result for EX. Combining the last two results,

$$\text{var}(X) = \frac{2}{\lambda^2} - \left(\frac{1}{\lambda}\right)^2 = \frac{1}{\lambda^2}$$

Example 5.6 **Power laws.** Let $\rho > 1$ and $f(x) = (\rho - 1)x^{-\rho}$ for $x \geq 1$.

$$EX = \int_1^\infty x(\rho - 1)x^{-\rho}\,dx = \frac{\rho - 1}{2 - \rho} x^{2-\rho}\Big|_1^\infty = \frac{\rho - 1}{\rho - 2}$$

if $\rho > 2$. $EX = \infty$ if $1 < \rho \leq 2$. The second moment

$$E(X^2) = \int_1^\infty x^2(\rho - 1)x^{-\rho}\,dx = \frac{\rho - 1}{3 - \rho} x^{3-\rho}\Big|_1^\infty = \frac{\rho - 1}{\rho - 3}$$

if $\rho > 3$. $E(X^2) = \infty$ if $1 < \rho \leq 2$. If $\rho > 3$,

$$\begin{aligned} \text{var}(X) &= \frac{\rho - 1}{\rho - 3} - \left(\frac{\rho - 1}{\rho - 2}\right)^2 \\ &= (\rho - 1)\left[\frac{(\rho - 2)^2 - (\rho - 1)(\rho - 3)}{(\rho - 2)^2(\rho - 3)}\right] = \frac{\rho - 1}{(\rho - 2)^2(\rho - 3)} \end{aligned}$$

If, for example, $\rho = 4$, $EX = 3/2$, $E(X^2) = 3$, and $\text{var}(X) = 3 - (3/2)^2 = 3/4$.

5.2 Distribution functions

Any random variable (discrete, continuous, or in between) has a *distribution function* defined by $F(x) = P(X \leq x)$. If X has a density function $f(x)$, then

$$F(x) = P(-\infty < X \leq x) = \int_{-\infty}^{x} f(y)\,dy$$

That is, F is an antiderivative of f, and a special one $F(x) \to 0$ as $x \to -\infty$ and $F(x) \to 1$ as $x \to \infty$.

One of the reasons for computing the distribution function is explained by the next formula. If $a < b$, then $\{X \leq b\} = \{X \leq a\} \cup \{a < X \leq b\}$ with the two sets on the right-hand side disjoint, so

$$P(X \leq b) = P(X \leq a) + P(a < X \leq b)$$

or, rearranging,

$$P(a < X \leq b) = P(X \leq b) - P(X \leq a) = F(b) - F(a) \tag{5.8}$$

The last formula is valid for any random variable. When X has density function f, it says that

$$\int_a^b f(x)\,dx = F(b) - F(a)$$

that is, the integral can be evaluated by taking the difference of any antiderivative at the two endpoints.

To see what distribution functions look like and to explain the use of (5.8), we return to our examples.

Example 5.7

Uniform distribution. $f(x) = 1/(b-a)$ for $a \le x \le b$.

$$F(x) = \begin{cases} 0 & x \le a \\ (x-a)/(b-a) & a \le x \le b \\ 1 & x \ge b \end{cases}$$

To check this, note that $P(a < X < b) = 1$, so $P(X \le x) = 1$ when $x \ge b$ and $P(X \le x) = 0$ when $x \le a$. For $a \le x \le b$, we compute

$$P(X \le x) = \int_{-\infty}^{x} f(y)\,dy = \int_a^x \frac{1}{b-a}\,dy = \frac{x-a}{b-a}$$

In the most important special case when $a = 0$ and $b = 1$, we have $F(x) = x$ for $0 \le x \le 1$.

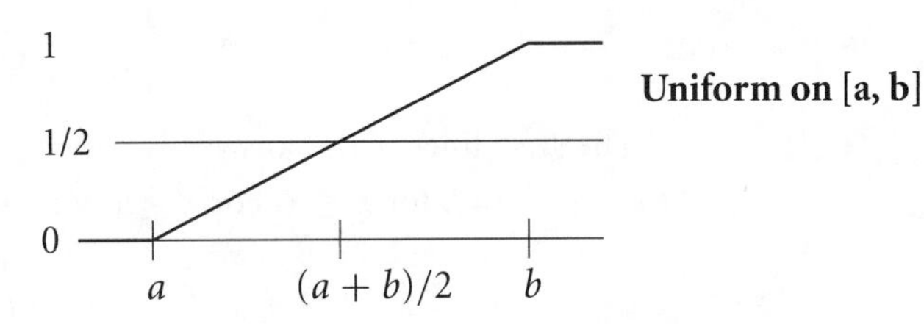

Uniform on [a, b]

Example 5.8

Exponential distribution. $f(x) = \lambda e^{-\lambda x}$ for $x \ge 0$.

$$F(x) = \begin{cases} 0 & x \le 0 \\ 1 - e^{-\lambda x} & x \ge 0 \end{cases}$$

The first line of the answer is easy to see. Since $P(X > 0) = 1$, we have $P(X \le x) = 0$ for $x \le 0$. For $x \ge 0$, we compute

$$P(X \le x) = \int_{-\infty}^{x} f(y)\,dy = \int_{0}^{x} \lambda e^{-\lambda y}\,dy$$
$$= -e^{-\lambda y}\Big|_0^x = -e^{-\lambda x} - (-1)$$

Suppose X has an exponential distribution with parameter λ. If $t \ge 0$, then $P(X > t) = 1 - P(X \le t) = 1 - F(t) = e^{-\lambda t}$, so if $s \ge 0$, then

$$P(T > t + s | T > t) = \frac{P(T > t + s)}{P(T > t)} = \frac{e^{-\lambda(t+s)}}{e^{-\lambda t}} = e^{-\lambda s} = P(T > s)$$

This is the **lack of memory property** of the exponential distribution. Given that you have been waiting t units of time, the probability that you must wait an additional s units of time is the same as if you had not been waiting at all.

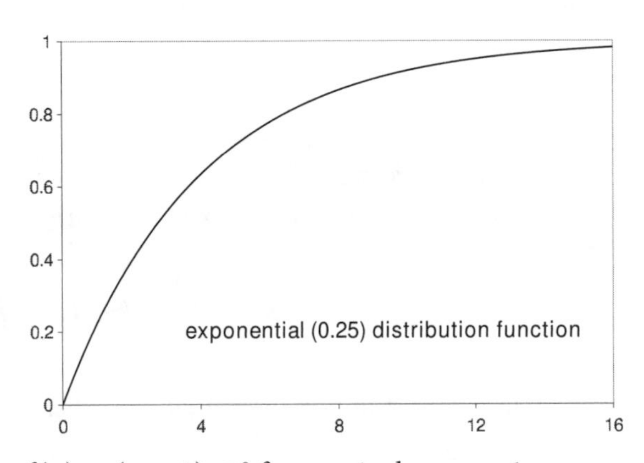

Example 5.9 **Power laws.** $f(x) = (\rho - 1)x^{-\rho}$ for $x \ge 1$ where $\rho > 1$.

$$F(x) = \begin{cases} 0 & x \le 1 \\ 1 - x^{-(\rho-1)} & x \ge 1 \end{cases}$$

The first line of the answer is easy to see. Since $P(X > 1) = 1$, we have $P(X \le x) = 0$ for $x \le 1$. For $x \ge 1$, we compute

$$P(X \le x) = \int_{-\infty}^{x} f(y)\,dy = \int_{1}^{x} (\rho - 1)y^{-\rho}\,dy$$
$$= -y^{-(\rho-1)}\Big|_1^x = 1 - x^{-(\rho-1)}$$

To illustrate the use of (5.8) we note that if $\rho = 3$, then

$$P(2 < X \le 4) = (1 - 4^{-2}) - (1 - 2^{-2}) = \frac{1}{4} - \frac{1}{16} = \frac{3}{16}$$

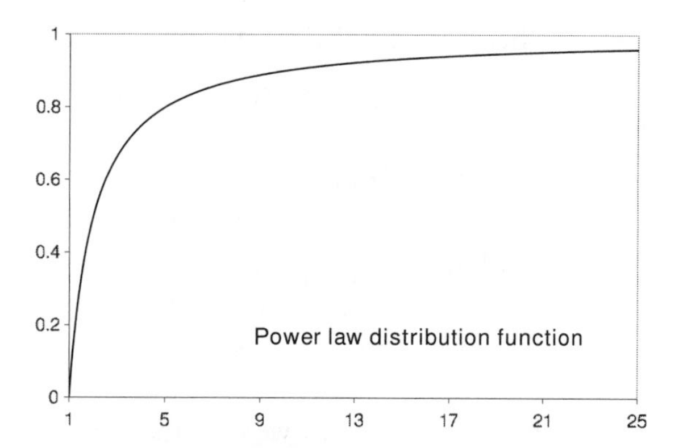

Distribution functions are somewhat messier in the discrete case.

Example 5.10 **Binomial(3, 1/2).** Flip three coins and let X be the number of heads that we see. The probability function is given by

x	0	1	2	3
$P(X = x)$	1/8	3/8	3/8	1/8

7/8

1/2 **Binomial(3, 1/2)**

1/8

0 1 2 3

In this case the distribution function is

$$F(x) = \begin{cases} 0 & x < 0 \\ 1/8 & 0 \le x < 1 \\ 1/2 & 1 \le x < 2 \\ 7/8 & 2 \le x < 3 \\ 1 & 3 \le x \end{cases}$$

To check this, note, for example, that for $1 \le x < 2$, $P(X \le x) = P(X \in \{0, 1\}) = 1/8 + 3/8$. The reader should note that F is discontinuous at each possible value of X and the height of the jump there is $P(X = x)$. The little black dots in the figure are there to indicate that at 0 the value is 1/8, at 1 it is 1/2, etc.

Theorem 5.1. *All distribution functions have the following properties:*

(i) *If* $x_1 < x_2$, *then* $F(x_1) \le F(x_2)$; *that is,* F *is nondecreasing.*
(ii) $\lim_{x \to -\infty} F(x) = 0$.
(iii) $\lim_{x \to \infty} F(x) = 1$.
(iv) $\lim_{y \downarrow x} F(y) = F(x)$; *that is,* F *is continuous from the right.*
(v) $\lim_{y \uparrow x} F(y) = P(X < x)$.
(vi) $\lim_{y \downarrow x} F(y) - \lim_{y \uparrow x} F(y) = P(X = x)$; *that is, the jump in* F *at* x *is equal to* $P(X = x)$.

Proof. To prove (i), we note that $\{X \le x_1\} \subset \{X \le x_2\}$, so (1.4) implies that $F(x_1) = P(X \le x_1) \le P(X \le x_2) = F(x_2)$.

To prove (ii), we note that $\{X \le x\} \downarrow \emptyset$ as $x \downarrow -\infty$ (here $\downarrow$ is short for "decreases and converges to"), so (1.5) implies that $P(X \le x) \downarrow P(\emptyset) = 0$.

The argument for (iii) is similar: $\{X \le x\} \uparrow \Omega$ as $x \uparrow \infty$ (here $\uparrow$ is short for "increases and converges to"), so (1.5) implies that $P(X \le x) \uparrow P(\Omega) = 1$.

To prove (iv), we note that if $y \downarrow x$, then $\{X \le y\} \downarrow \{X \le x\}$, so (1.5) implies that $P(X \le y) \downarrow P(X \le x)$.

The argument for (v) is similar. If $y \uparrow x$, then $\{X \le y\} \uparrow \{X < x\}$ since $\{X = x\} \not\subset \{X \le y\}$ when $y < x$. Using (1.5) now, (v) follows.

Subtracting (v) from (iv) gives (vi). □

5.2.1 Two useful transformations

The first result can often be used to reduce a general continuous distribution to the special case of a uniform.

Theorem 5.2. *Suppose* X *has a continuous distribution. Then* $Y = F(X)$ *is uniform on* $(0, 1)$.

Proof. Even though F may not be strictly increasing, we can define an inverse of F by

$$F^{-1}(y) = \min\{x\colon F(x) \ge y\}$$

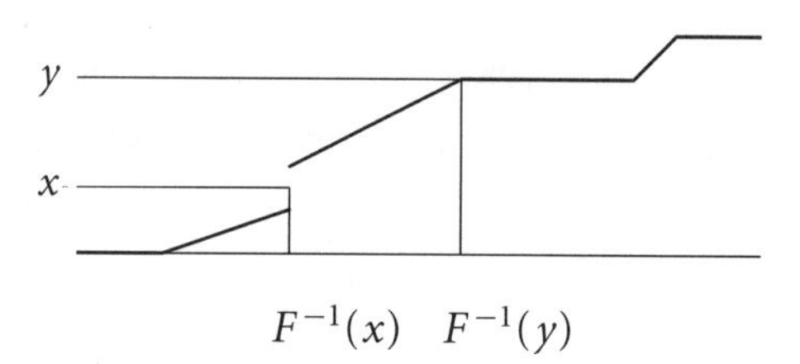

Using this definition of F^{-1}, we have

$$P(F(X) < y) = P(X < F^{-1}(y)) = F(F^{-1}(y)) = y$$

the last equality holding since F is continuous. □

This is the key to many results in nonparametric statistics. For example, suppose we have a sample of 10 men's heights and 10 women's heights. To test the hypothesis that men and women have the same height distribution, we can look at the ranks of the men's heights in the overall sample of size 20. For example, these might be 1, 2, 3, 4, 6, 8, 9, 11, 13, and 14. Since applying the distribution function to the data points does not change the ranks, Theorem 5.2 implies that the distribution of the rank sum does not depend on the underlying distribution.

Reversing the ideas in the proof of Theorem 5.2, we get a result that is useful to construct random variables with a specified distribution.

Theorem 5.3. *Suppose U has a uniform distribution on (0, 1). Then $Y = F^{-1}(U)$ has distribution function F.*

Proof. The definition of F^{-1} was chosen so that if $0 < x < 1$, then

$$F^{-1}(y) \leq x \text{ if and only if } F(x) \geq y$$

and this holds for any distribution function F. Taking $y = U$, it follows that

$$P(F^{-1}(U) \leq x) = P(U \leq F(x)) = F(x)$$

since $P(U \leq u) = u$. □

For a concrete example, suppose we want to construct an exponential distribution with parameter λ. Setting $1 - e^{-\lambda x} = y$ and solving gives $-\ln(1 - y)/\lambda = x$. So if U is uniform on (0, 1), then $-\ln(1 - U)/\lambda$ has the desired exponential distribution. Of course since $1 - U$ is uniform on (0, 1), we could also use $-\ln(U)/\lambda$. In the case of a power law, setting $1 - x^{-(\rho-1)} = y$ and solving gives $(1 - y)^{-1/(\rho-1)} = x$. So if U is uniform on (0, 1), then $U^{-1/(\rho-1)}$ has the desired power law distribution.

5.2.2 Medians

Intuitively, the median is the place where $F(x)$ crosses 1/2. The precise definition we are about to give is complicated by the fact that $\{x\colon F(x) = 1/2\}$ may be empty or contain more than one point.

m is a **median** for F if $P(X \leq m) \geq 1/2$ and $P(X \geq m) \geq 1/2$.

We begin with our three favorite examples.

Example 5.11

Uniform distribution. Suppose X has density $1/(b-a)$ for $a \le x \le b$. As we computed in Example 5.7, the distribution function is $(x-a)/(b-a)$ for $a \le x \le b$. The computation of the median is illustrated in Example 5.7. To find the median, we set $(x-a)/(b-a) = 1/2$, that is, $2x - 2a = b - a$, or solving we have $x = (b+a)/2$. To see that this is the only median, we observe that if $m < (a+b)/2$, then $P(X \le m) < 1/2$, while if $m > (a+b)/2$, then $P(X \ge m) < 1/2$. In this case the median is equal to the mean, but this is a rare occurrence.

Example 5.12

Exponential distribution. Suppose X has density $\lambda e^{-\lambda x}$ for $x \ge 0$. As we computed in Example 5.8, the distribution function is $F(x) = 1 - e^{-\lambda x}$. To find the median, we set $P(X \le m) = 1/2$, that is, $1 - e^{-\lambda m} = 1/2$, and solve to find $m = (\ln 2)/\lambda$, compared to the mean $1/\lambda$.

In the context of radioactive decay, which is commonly modeled with an exponential distribution, the median is sometimes called the **half-life**, since half the particles will have broken down by that time. One reason for interest in the half-life is that

$$P(X > k \ln 2/\lambda) = e^{-k \ln 2} = 2^{-k}$$

or in words, after k half-lives only $1/2^k$ particles remain radioactive.

Example 5.13

Power laws. Suppose X has density $(\rho - 1)x^{-\rho}$ for $x \ge 1$, where $\rho > 1$. As we computed in Example 5.9, the distribution function is $1 - x^{-(\rho-1)}$. To find the median, we set $1 - m^{-(\rho-1)} = 1/2$, that is, $1/2 = m^{-(\rho-1)}$, and solving gives $m = 2^{1/\rho-1}$. This contrasts to the mean $(\rho-1)/(\rho-2)$, which is finite for $\rho > 2$. For a concrete example note that when $\rho = 4$, the mean is 3/2 while the median is $2^{1/4} = 1.189$.

We now turn to unusual cases where there may be no solution to $P(X \le x) = 1/2$ or more than 1.

Example 5.14

Multiple solutions: Binomial(3, 1/2). The distribution function was computed in Example 5.10.

$$F(x) = \begin{cases} 0 & x < 0 \\ 1/8 & 0 \le x < 1 \\ 1/2 & 1 \le x < 2 \\ 7/8 & 2 \le x < 3 \\ 1 & 3 \le x \end{cases}$$

If $1 \le m \le 2$, then $P(X \le m) \ge 1/2$ and $P(X \ge m) \ge 1/2$, so the set of medians is $[1, 2]$. For a picture see Example 5.10.

Example 5.15

No solution: Uniform on 1, 2, 3. Suppose X takes values 1, 2, 3 with probability 1/3 each. The distribution function is

$$F(x) = \begin{cases} 0 & x < 1 \\ 1/3 & 1 \le x < 2 \\ 2/3 & 2 \le x < 3 \\ 1 & 3 \le x \end{cases}$$

To check that 2 is a median, we note that

$$P(X \le 2) = P(X \in \{1, 2\}) = 2/3$$
$$P(X \ge 2) = P(X \in \{2, 3\}) = 2/3$$

This is the only median since if $x < 2$, then $P(X \le x) \le P(X < 2) \le 1/3$ and if $x > 2$, then $P(X \ge x) \le P(X > 2) = 1/3$.

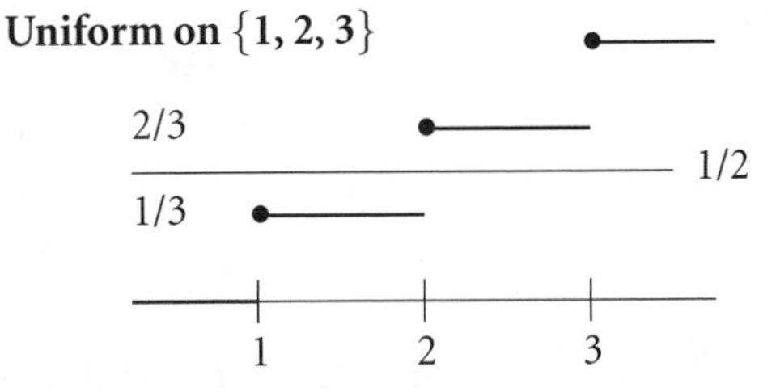

5.3 Functions of random variables

In this section we answer the question, "If X has density function f and $Y = r(X)$, then what is the density function for Y?" Before proving a general result, we consider an example.

Example 5.16

Suppose X has an exponential distribution with parameter λ. What is the distribution of $Y = X^2$?

To solve this problem we will use the distribution function. First, we recall from Example 5.8 that $P(X \le x) = 1 - e^{-\lambda x}$, so if $y \ge 0$, then

$$P(Y \le y) = P(X^2 \le y) = P(X \le \sqrt{y}) = 1 - e^{-\lambda y^{1/2}}$$

Differentiating, we see that the density function of Y is given by

$$f_Y(y) = \frac{d}{dy} P(Y \le y) = \frac{\lambda y^{-1/2}}{2} e^{-\lambda y^{1/2}} \quad \text{for } y \ge 0$$

and 0 otherwise.

Generalizing from the last example, we get

Theorem 5.4. *Suppose X has density f and $P(a < X < b) = 1$. Let $Y = r(X)$. Suppose $r : (a, b) \to (\alpha, \beta)$ is continuous and strictly increasing, and let $s : (\alpha, \beta) \to (a, b)$ be the inverse of r. Then Y has density*

$$g(y) = f(s(y))s'(y) \quad \textit{for } y \in (\alpha, \beta) \tag{5.9}$$

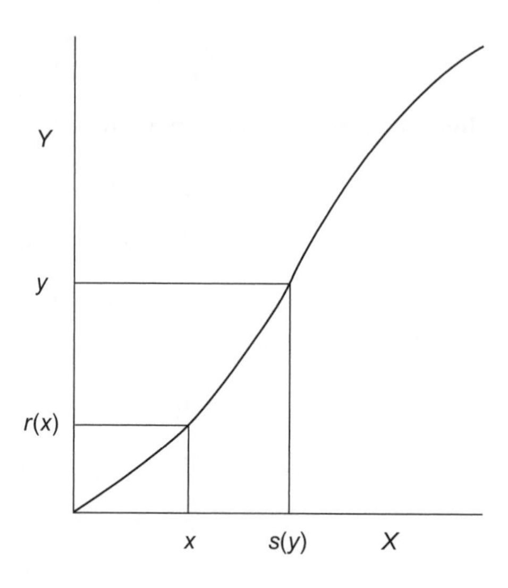

Before proving this, let's see how it applies to the last example. There X has density $f(x) = \lambda e^{-\lambda x}$ for $x \geq 0$, so we can take $a = 0$ and $b = \infty$. The function $r(x) = x^2$ is indeed continuous and strictly increasing on $(0, \infty)$. To find the inverse function, we set $y = x^2$ and solve to get $x = y^{1/2}$, so $s(y) = y^{1/2}$. Differentiating, we have $s'(y) = y^{-1/2}/2$, and plugging into the formula, we have

$$g(y) = \lambda e^{-\lambda y^{1/2}} \cdot y^{-1/2}/2 \quad \text{for } y > 0$$

Proof. If $y \in (\alpha, \beta)$, then

$$P(Y \leq y) = P(r(X) \leq y) = P(X \leq s(y))$$

since r is increasing and s is its inverse. Writing $F(x)$ for $P(X \leq x)$ and differentiating with respect to y now gives

$$g(y) = \frac{d}{dy} P(Y \leq y) = \frac{d}{dy} F(s(y)) = F'(s(y))s'(y) = f(s(y))s'(y)$$

by the chain rule. □

For our next example, we will consider a special case of Theorem 5.2. If X has a continuous distribution function F then $F(X)$ is uniform on $(0, 1)$.

Example 5.17

Exponential to uniform. Suppose X has an exponential distribution with parameter 3. That is, X has density function $3e^{-3x}$ for $x \geq 0$. Find the distribution function of $Y = 1 - e^{-3X}$.

Here, $r(x) = 1 - e^{-3x}$ is increasing on $(0, \infty)$, $\alpha = r(0) = 0$, and $\beta = r(\infty) = 1$. To find the inverse function, we set $y = 1 - e^{-3x}$ and solve to get $s(y) = (-1/3)\ln(1 - y)$. Differentiating, we have $s'(y) = -(-1/3)/(1 - y)$. So plugging into (3.1), the density function of Y is

$$f(s(y))s'(y) = 3e^{\ln(1-y)} \cdot \frac{1/3}{(1-y)} = 1$$

for $0 < y < 1$. That is, Y is uniform on $(0, 1)$.

Example 5.18

Cauchy distribution. A drunk standing 1 foot from a wall shines a flashlight at a random angle Θ that is uniformly distributed between $-\pi/2$ and $\pi/2$. Find the density function of the place X where the light hits the wall.

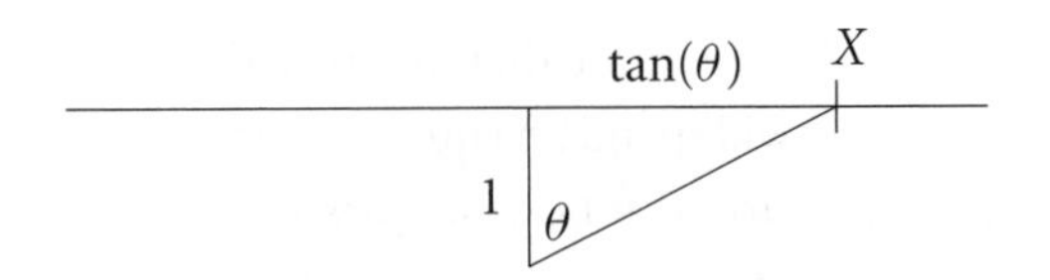

The angle Θ is uniformly distributed on $[-\pi/2, \pi/2]$ and has density $f(\theta) = 1/\pi$. As you can see from the picture, $r(\theta) = \tan(\theta)$. The inverse function $s(x) = \tan^{-1}(x)$ has $s'(x) = 1/(1 + x)$, so using (5.9) X has density function

$$\frac{1}{\pi} \cdot \frac{1}{1 + x^2}$$

This is the Cauchy density. Its median is 0 but its mean does not exist since

$$E|X| = \int \frac{|x|}{\pi(1 + x^2)} \, dx = \infty$$

To check the last conclusion note that the integrand is $\approx 1/|x|$ when $|x|$ is large.

Example 5.19

How not to water your lawn. The head of a lawn sprinkler, which is a metal rod with a line of small holes in it, revolves back and forth so that drops of water shoot out at angles between 0 and $\pi/2$ radians (that is, between 0 and 90°). If we use x to denote the distance from the sprinkler and y the height off the ground then a drop of water released at angle θ with velocity v_0 will follow a trajectory

$$x(t) = (v_0 \cos\theta)t \qquad y(t) = (v_0 \sin\theta)t - gt^2/2$$

where g is the gravitational constant, 32 ft/s^2. The drop lands when $y(t_0) = 0$, that is, at time $t_0 = (2v_0 \sin\theta)/g$. At this time

$$x(t_0) = \frac{2v_0^2}{g} \sin\theta \cos\theta = \frac{v_0^2}{g} \sin(2\theta)$$

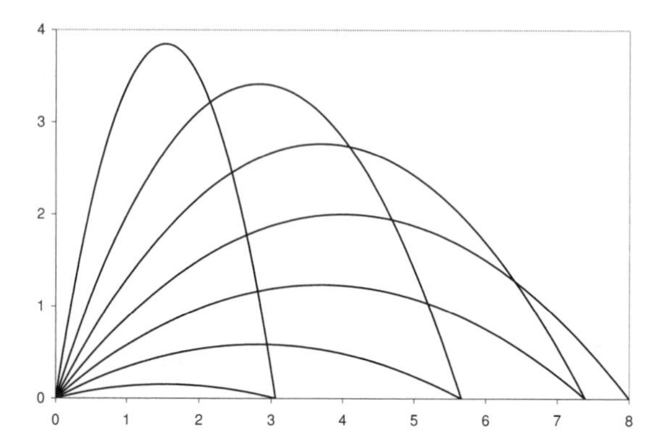

If we assume that the sprinkler moves evenly back and forth between 0 and $\pi/2$, it will spend an equal amount of time at each angle. Letting $K = v_0^2/g$, this leads us to the following question:

> If Θ is uniform on $[0, \pi/2]$, then what is the distribution of $Z = K \sin(2\Theta)$?

The first difficulty we must confront when solving this problem is that $\sin(2x)$ is increasing on $[0, \pi/4]$ and decreasing on $[\pi/4, \pi/2]$. The solution to this problem is simple, however. The function $\sin(2x)$ is symmetric about $\pi/4$, so if we let X to be uniform on $[0, \pi/4]$, then $Z = K \sin(2\Theta)$ and $Y = K \sin(2X)$ have the same distribution. To apply (5.9), we let $r(x) = K \sin(2x)$ and solve $y = K \sin(2x)$ to get $s(y) = (1/2) \sin^{-1}(y/K)$. Plugging into (5.9) and recalling

$$\frac{d}{dx} \sin^{-1}(x) = \frac{1}{\sqrt{1 - x^2}}$$

we see that Y has density function

$$f(s(y))s'(y) = \frac{4}{\pi} \cdot \frac{1}{2\sqrt{1 - y^2/K^2}} \cdot \frac{1}{K} = \frac{2}{\pi\sqrt{K^2 - y^2}}$$

when $0 < y < K$ and 0 otherwise. The title of this example comes from the fact that the density function goes to ∞ as $y \to K$, so the lawn gets very soggy at the edge of the sprinkler's range. This is due to the fact that $s'(K) = \infty$, which in turn is caused by $r'(\pi/4) = 0$.

5.4 Joint distributions

Two random variables are said to have **joint density function** f if for any $A \subset \mathbf{R}^2$,

$$P((X, Y) \in A) = \iint_A f(x, y)\, dx\, dy \tag{5.10}$$

where $f(x, y) \geq 0$ and $\iint f(x, y)\, dx\, dy = 1$.

In words, we find the probability that (X, Y) lies in A by integrating f over A. As we will see a number of times later, it is useful to think of $f(x, y)$ as $P(X = x, Y = y)$ even though the last event has probability 0. As in Section 5.1, the precise interpretation of $f(x, y)$ is

$$\begin{aligned} P(x \leq X \leq x + \Delta x,\ y \leq Y \leq y + \Delta y) &= \int_x^{x+\Delta x} \int_y^{y+\Delta y} f(u, v)\, dv\, du \\ &\approx f(x, y)\Delta x \Delta y \end{aligned}$$

when Δx and Δy are small, so $f(x, y)$ indicates how likely it is for (X, Y) to be near (x, y).

We are now going to give three concrete examples of a joint density function. As in the case of density functions we will only give the formula for $f(x)$ where it is positive and omit the phrase "0 otherwise."

Example 5.20 $f(x, y) = e^{-y} \quad 0 < x < y < \infty$.

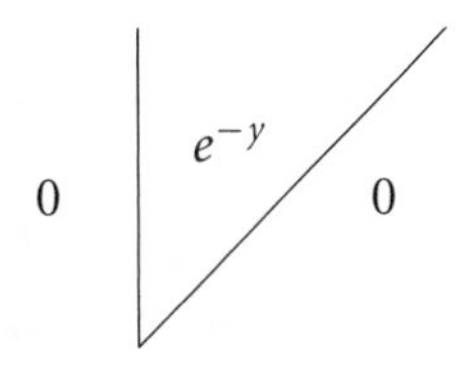

The story behind this example is told later in Example 5.28. To check that f is a density function, we observe that

$$\int_0^\infty \int_0^y e^{-y}\, dx\, dy = \int_0^\infty y e^{-y}\, dy$$

and integrating by parts (5.7) with $g(y) = y$, $h'(y) = e^{-y}$, so $g'(y) = 1$ and $h(y) = -e^{-y}$.

$$\int_0^\infty y e^{-y}\, dy = -y e^{-y}\big|_0^\infty + \int_0^\infty e^{-y} dy = 0 + (-e^{-y})\big|_0^\infty = 1$$

To illustrate the use of (5.10) we will now compute $P(X \le 1)$, which can be written as $P((X, Y) \in A)$, where $A = \{(x, y) : x \le 1\}$. The formula in (5.10) tells us that we find $P((X, Y) \in A)$ by integrating the joint density over A. However, the joint density is positive only on $B = \{(x, y) : 0 < x < y < \infty\}$, so we need to integrate only over $A \cap B = \{(x, y) : 0 < x \le 1, x < y\}$, and doing this we find

$$P(X \le 1) = \int_0^1 \int_x^\infty e^{-y}\, dy\, dx$$

To evaluate the double integral we begin by observing that

$$\int_x^\infty e^{-y}\, dy = (-e^{-y})\big|_x^\infty = 0 - (-e^{-x}) = e^{-x}$$

so $P(X < 1) = \int_0^1 e^{-x}\, dx = (-e^{-x})\big|_0^1 = 1 - e^{-1}$.

Example 5.21

Uniform distribution on a ball. Pick a point at random from the ball $B = \{(x, y): x^2 + y^2 \le 1\}$. By "at random from B" we mean that a choice outside of B is impossible and that all the points in B should be equally likely. In terms of the joint density this means that $f(x, y) = 0$ when $(x, y) \notin B$ and there is a constant $c > 0$ so that $f(x, y) = c$ when $(x, y) \in B$.

Our $f(x, y) \ge 0$. To make the integral of f equal to 1, we have to choose c appropriately. Now,

$$\iint f(x, y)\, dx\, dy = \iint_B c\, dx\, dy = c\,(\text{area of } B) = c\pi$$

So we choose $c = 1/\pi$ to make the integral 1 and define

$$f(x, y) = 1/\pi \quad x^2 + y^2 \le 1$$

The arguments that led to the last conclusion generalize easily to show that if we pick a point "at random" from a set S with area a, then

$$f(x, y) = \frac{1}{a} \quad (x, y) \in S \tag{5.11}$$

Example 5.22

Buffon's needle. A floor consists of boards of width 1. If we drop a needle of length $L \le 1$ on the floor, what is the probability it will touch one of the cracks (that is, the small spaces between the boards)? To make the question simpler to answer, we assume that the needle and the cracks have width zero.

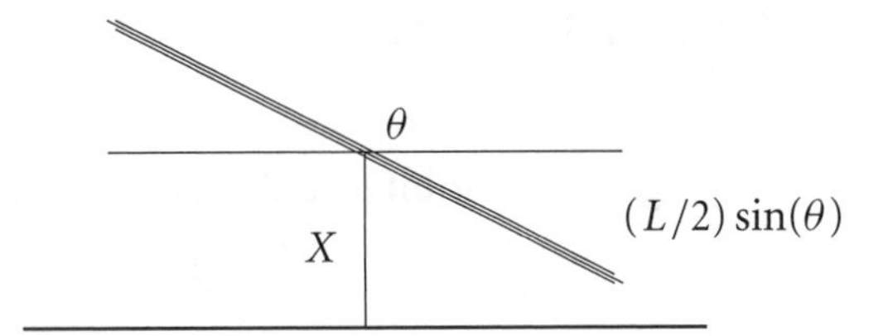

Let X be the distance from the center of the needle to the nearest crack and Θ be the angle $\in [0, \pi)$ that the top half of the needle makes with the crack. (We make this choice to have $\sin \Theta > 0$.) We assume that all the ways the needle can land are equally likely; that is, the joint distribution of (X, Θ) is

$$f(x, \theta) = \frac{2}{\pi} \quad \text{if } x \in [0, 1/2), \quad \theta \in [0, \pi)$$

The formula for the joint density follows from (5.11). We are picking a point "at random" from a set S with area $\pi/2$, so the joint density is $2/\pi$ on S.

By drawing a picture (like the one given earlier), one sees that the needle touches the crack if and only if $(L/2) \sin \Theta \geq X$. (5.10) tells us that the probability of this event is obtained by integrating the joint density over

$$A = \{(x, \theta) \in [0, 1/2) \times [0, \pi): x \leq (L/2) \sin \theta\}$$

so the probability we seek is

$$\begin{aligned}\iint_A f(x, \theta)\, dx\, d\theta &= \int_0^\pi \int_0^{(L/2)\sin\theta} \frac{2}{\pi}\, dx\, d\theta \\ &= \frac{2}{\pi} \int_0^\pi \frac{L}{2} \sin\theta\, d\theta = \frac{L}{\pi}(-\cos\theta)\Big|_0^\pi = 2L/\pi\end{aligned}$$

Buffon wanted to use this as a method of estimating π. Taking $L = 1/2$ and performing the experiment 10,000 times on a computer, we found that 1 over the fraction of times the needle hit the crack was 3.2310, 3.1368, and 3.0893 in the three times we tried this. We show in Chapter 6 that these numbers are typical outcomes and that to compute π to four decimal places would require about 10^8 (or 100 million) tosses.

Remark. Before leaving the subject of joint densities, we would like to make one remark that will be useful later. If X and Y have joint density $f(x, y)$, then $P(X = Y) = 0$. To see this, we observe that $\iint_A f(x, y)\, dx\, dy$ is the volume of the region over A underneath the graph of f, but this volume is 0 if A is the line $x = y$.

5.4.1 Joint distribution functions

The joint distribution of two random variables is occasionally described by giving the **joint distribution function**:

$$F(x, y) = P(X \leq x, Y \leq y)$$

The next example illustrates this notion but also shows that sometimes the density function is easier to write down.

Example 5.23

Suppose (X, Y) is uniformly distributed over the square $\{(x, y) : 0 < x < 1, 0 < y < 1\}$. That is,

$$f(x, y) = 1 \quad 0 < x, y < 1$$

Here, we are picking a point "at random" from a set with area 1, so the formula follows from (5.11).

By patiently considering the possible cases, we find that

$$F(x, y) = \begin{cases} 0 & \text{if } x < 0 \text{ or } y < 0 \\ xy & \text{if } 0 \leq x \leq 1 \text{ and } 0 \leq y \leq 1 \\ x & \text{if } 0 \leq x \leq 1 \text{ and } y > 1 \\ y & \text{if } x > 1 \text{ and } 0 \leq y \leq 1 \\ 1 & \text{if } x > 1 \text{ and } y > 1 \end{cases}$$

The answer is probably easier to understand in a picture:

	$x < 0$	$0 \le x \le 1$	$x > 1$
$y > 1$	0	x	1
$0 \le y \le 1$	0	xy	y
$y < 0$	0	0	0

(Horizontal lines at $y = 0$ and $y = 1$; vertical lines at $x = 0$ and $x = 1$.)

The first case should be clear: If $x < 0$ or $y < 0$, then $\{X \leq x, Y \leq y\}$ is impossible since X and Y always lie between 0 and 1. For the second case we note that when $0 \leq x \leq 1$ and $0 \leq y \leq 1$,

$$P(X \leq x, Y \leq y) = \int_0^x \int_0^y 1 \, dv \, du = xy$$

In the third case, since values of $Y > 1$ are impossible,

$$P(X \leq x, Y \leq y) = P(X \leq x, Y \leq 1) = x$$

by the formula for the second case. The fourth case is similar to the third, and the fifth is trivial. X and Y are always smaller than 1, so if $x > 1$ and $y > 1$, then $\{X \le x, Y \le y\}$ has probability 1.

We will not use the joint distribution function in what follows. For completeness, however, we want to mention two of its important properties. The first formula is the two-dimensional generalization of $P(a < X \le b) = F(b) - F(a)$.

$$\begin{aligned} &P(a_1 < X \le b_1, a_2 < Y \le b_2) \\ &\quad = F(b_1, b_2) - F(a_1, b_2) - F(b_1, a_2) + F(a_1, a_2) \end{aligned} \tag{5.12}$$

Proof. The reasoning we use here is much like that employed in studying the probabilities of unions in Section 1.6. By adding and subtracting the probabilities on the right, we end up with the desired area counted exactly once.

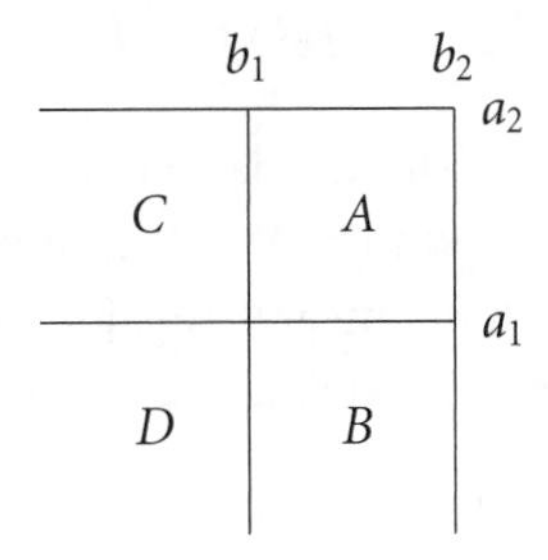

Using A as shorthand for $P((X, Y) \in A)$, etc., and consulting the picture,

$$\begin{array}{rcllll} F(b_1, b_2) & = & A & +B & +C & +D \\ -F(a_1, b_2) & = & & -B & & -D \\ -F(b_1, a_2) & = & & & -C & -D \\ F(a_1, a_2) & = & & & & +D \end{array}$$

Adding the last four equations gives the one in (5.12). □

The next formula tells us how to recover the joint density function from the joint distribution function. To motivate the formula, we recall that in one dimension $F' = f$ since $F(x) = \int_{\infty}^{x} f(u)\,du$.

$$\frac{\partial^2 F}{\partial x \partial y} = f \tag{5.13}$$

To explain why this formula is true, we note that

$$F(x, y) = \int_{-\infty}^{x} \int_{-\infty}^{y} f(u, v)\,dv\,du$$

and differentiating twice kills the two integrals. To check that (5.13) works in Example 5.23, $F(x, y) = xy$ when $0 < x < 1$ and $0 < y < 1$, so $\frac{\partial^2 F}{\partial x \partial y} = 1$ there and it is 0 otherwise.

5.5 Marginal and conditional distributions

In the discrete case the marginal distributions are obtained from the joint distribution by summing

$$P(X = x) = \sum_y P(X = x, Y = y) \qquad P(Y = y) = \sum_x P(X = x, Y = y)$$

In the continuous case if X and Y have joint density $f(x, y)$, then the **marginal densities** of X and Y are given by

$$f_X(x) = \int f(x, y)\, dy \qquad f_Y(y) = \int f(x, y)\, dx \tag{5.14}$$

The verbal explanation of the first formula is similar to that of the discrete case: if $X = x$, then Y will take on some value y, so to find the marginal density $f_X(x)$, we integrate the joint density $f(x, y)$ over all possible values of y.

To illustrate the use of these formulas we look at Example 5.20.

Example 5.24 $f(x, y) = e^{-y} \quad 0 \le x \le y < \infty$.

In this case

$$f_X(x) = \int_x^\infty e^{-y}\, dy = (-e^{-y})\big|_x^\infty = e^{-x}$$

since (5.14) tells us to integrate $f(x, y)$ over all values of y but we only have $f > 0$ when $y > x$. Similarly,

$$f_Y(y) = \int_0^y e^{-y}\, dx = ye^{-y}$$

Two random variables with joint density f are **independent** if and only if

$$f(x, y) = f_X(x) f_Y(y)$$

that is, if the joint density is the product of the marginal densities.

We will now consider three examples that parallel the ones used in the discrete case.

Example 5.25 $f(x, y) = e^{-y} \quad 0 \le x \le y < \infty$.

We calculated the joint distribution in the previous example but we can settle the question without computation. $f(3, 2) = 0$ while $f_X(3)$ and $f_Y(2)$ are both positive, so

$$f(3,2) = 0 < f_X(3) f_Y(2)$$

and X and Y are not independent. In general, if the set of values where $f > 0$ is not a rectangle then X and Y are not independent.

Example 5.26 $f(x,y) = (1+x+y)/2 \quad 0 \le x, y \le 1.$

In this case the set where $f > 0$ is a rectangle, so the joint distribution passes the first test and we have to compute the marginal densities

$$f_X(x) = \int_0^1 (1+x+y)/2\, dy = \left(\frac{1+x}{2}\right) y + \frac{y^2}{4}\Big|_0^1 = \frac{x}{2} + \frac{3}{4}$$

$$f_Y(y) = \frac{y}{2} + \frac{3}{4} \quad \text{by symmetry}$$

These formulas are valid for $0 \le x \le 1$ and $0 \le y \le 1$ respectively. To check independence we have to see if

$$\frac{1+x+y}{2} = \left(\frac{x}{2} + \frac{3}{4}\right) \cdot \left(\frac{y}{2} + \frac{3}{4}\right) \qquad (\star)$$

A simple way to see that $(\star)$ is wrong is simply to note that when $x = y = 0$, it says that $1/2 = 9/16$.

Example 5.27

$$f(x,y) = \frac{y^{-3/2}\cos x}{(e-1)} e^{\sin x - (2/\sqrt{y})} \quad 0 < x < \pi/2,\ y > 0$$

In this case, the integration does not look like much fun, so we adopt another approach.

Theorem 5.5. *If the joint density function $f(x,y)$ can be written as $g(x)h(y)$, then there is a constant c so that $f_X(x) = cg(x)$ and $f_Y(y) = h(y)/c$. It follows that $f(x,y) = f_X(x) f_Y(y)$ and hence X and Y are independent.*

In words, if we can write $f(x,y)$ as a product of a function of x and a function of y then these functions must be constant multiples of the marginal densities. Theorem 5.5 takes care of our example since

$$f(x,y) = \left(\frac{\cos x\, e^{\sin x}}{e-1}\right)\left(y^{-3/2} e^{-2/\sqrt{y}}\right)$$

Proof. We begin by observing

$$f_X(x) = \int f(x, y)\,dy = g(x) \int h(y)\,dy$$

$$f_Y(y) = \int f(x, y)\,dx = h(y) \int g(x)\,dx$$

$$1 = \int\int f(x, y)\,dx\,dy = \int g(x)\,dx \int h(y)\,dy$$

So if we let $c = \int h(y)\,dy$, then the last equation implies $\int g(x)\,dx = 1/c$ and the first two give us $f_X(x) = cg(x)$ and $f_Y(y) = h(y)/c$. □

Conditional distributions

Introducing $f_X(x|Y = y)$ as notation for the **conditional density of** X **given** $Y = y$ (which we think of as $P(X = x|Y = y)$), we have

$$f_X(x|Y = y) = \frac{f(x, y)}{f_Y(y)} = \frac{f(x, y)}{\int f(u, y)\,du} \tag{5.15}$$

In words, we fix y, consider the joint density function as a function of x, and then divide by the integral to make it a probability density. To see how formula (5.15) works, we return to Example 5.15.

Example 5.28 $f(x, y) = e^{-y} \quad 0 \le x \le y < \infty$

In this case we have computed $f_Y(y) = ye^{-y}$ (in Example 5.24), so

$$f_X(x|Y = y) = \frac{e^{-y}}{ye^{-y}} = \frac{1}{y} \quad \text{for } 0 < x < y$$

That is, the conditional distribution is uniform on $(0, y)$. This should not be surprising since $x \to f(x, y)$ is constant for $0 < x < y$ and 0 otherwise.

To compute the other conditional distribution we recall $f_X(x) = e^{-x}$, so

$$f_Y(y|X = x) = \frac{e^{-y}}{e^{-x}} = e^{-(y-x)} \quad \text{for } y > x$$

That is, given $X = x$, $Y - x$ is exponential with parameter 1. From this it follows that if Z_1, Z_2 are independent exponential(1), then $X = Z_1$, $Y = Z_1 + Z_2$ has the joint distribution given previously. If we condition on $X = x$, then $Z_1 = x$ and $Y = x + Z_2$.

The multiplication rule says

$$P(X = x, Y = y) = P(X = x)P(Y = y|X = x)$$

Substituting in the analogous continuous quantities, we have

$$f(x, y) = f_X(x) f_Y(y|X = x) \tag{5.16}$$

The next example demonstrates the use of (5.16) to compute a joint distribution.

Example 5.29 Suppose we pick a point uniformly distributed on (0, 1), call it X, and then pick a point Y uniformly distributed on $(0, X)$.

To find the joint density of (X, Y) we note that

$$\begin{aligned} f_X(x) &= 1 && \text{for } 0 < x < 1 \\ f_Y(y|X = x) &= 1/x && \text{for } 0 < y < x \end{aligned}$$

So using (5.16), we have

$$f(x, y) = f_X(x) f_Y(y|X = x) = 1/x \quad \text{for } 0 < y < x < 1$$

To complete the picture we compute

$$f_Y(y) = \int f(x, y)\, dx = \int_y^1 \frac{1}{x}\, dx = -\ln y$$

$$f_X(x|Y = y) = \frac{f(x, y)}{f_Y(y)} = \frac{1/x}{-\ln y} \quad \text{for } y < x < 1$$

Again the conditional density of X given $Y = y$ is obtained by fixing y, regarding the joint density function as a function of x, and then normalizing so that the integral is 1. The reader should note that although X is uniform on (0, 1) and Y is uniform on $(0, X)$, X is not uniform on $(Y, 1)$ but has a greater probability of being near Y.

5.6 Exercises

Density functions

1. Suppose X has density function $f(x) = c(3 - |x|)$ when $-3 < x < 3$. What value of c makes this a density function?

2. Consider $f(x) = c(1 - x^2)$ for $-1 < x < 1$ and 0 otherwise. What value of c should we take to make f a density function?

3. Suppose X has density function $6x(1 - x)$ for $0 < x < 1$ and 0 otherwise. Find (a) EX, (b) $E(X^2)$, and (c) var (X).

4. Suppose X has density function $x^2/9$ for $0 < x < 3$ and 0 otherwise. Find (a) EX, (b) $E(X^2)$, and (c) $\text{var}(X)$.

5. Suppose X has density function $x^{-2/3}/21$ for $1 < x < 8$ and 0 otherwise. Find (a) EX, (b) $E(X^2)$, and (c) $\text{var}(X)$.

Distribution functions

6. $F(x) = 3x^2 - 2x^3$ for $0 < x < 1$ (with $F(x) = 0$ if $x \le 0$ and $F(x) = 1$ if $x \ge 1$) defines a distribution function. Find the corresponding density function.

7. Let $F(x) = e^{-1/x}$ for $x > 0$ and $F(x) = 0$ for $x \le 0$. Is F a distribution function? If so, find its density function.

8. Let $F(x) = 3x - 2x^2$ for $0 \le x \le 1$, $F(x) = 0$ for $x \le 0$, and $F(x) = 1$ for $x \ge 1$. Is F a distribution function? If so, find its density function.

9. Suppose X has density function $f(x) = x/2$ for $0 < x < 2$ and 0 otherwise. Find (a) the distribution function, (b) $P(X < 1)$, (c) $P(X > 3/2)$, and (d) the median.

10. Suppose X has density function $f(x) = 4x^3$ for $0 < x < 1$ and 0 otherwise. Find (a) the distribution function, (b) $P(X < 1/2)$, (c) $P(1/3 < X < 2/3)$, and (d) the median.

11. Suppose X has density function $x^{-1/2}/2$ for $0 < x < 1$ and 0 otherwise. Find (a) the distribution function, (b) $P(X > 3/4)$, (c) $P(1/9 < X < 1/4)$, and (d) the median.

12. Suppose $P(X = x) = x/21$ for $x = 1, 2, 3, 4, 5, 6$. Find all the medians of this distribution.

13. Suppose X has a Poisson distribution with $\lambda = \ln 2$. Find all the medians of X.

14. Suppose X has a geometric distribution with success probability 1/4; that is, $P(X = k) = (3/4)^{k-1}(1/4)$. Find all the medians of X.

15. Suppose X has density function $3x^{-4}$ for $x \ge 1$. (a) Find a function g so that $g(X)$ is uniform on $(0, 1)$. (b) Find a function h so that if U is uniform on $(0, 1)$, $h(U)$ has density function $3x^{-4}$ for $x \ge 1$.

16. Suppose $X_1, \ldots, X_n$ are independent and have distribution function $F(x)$. Find the distribution functions of (a) $Y = \max\{X_1, \ldots, X_n\}$ and (b) $Z = \min\{X_1, \ldots, X_n\}$

17. Suppose $X_1, \ldots, X_n$ are independent exponential(λ). Show that

$$\min\{X_1, \ldots, X_n\} = \text{exponential}(n\lambda)$$

Functions of random variables

18. Suppose X has density function $f(x)$ for $a \le x \le b$ and $Y = cX + d$, where $c > 0$. Find the density function of Y.

19. Show that if $X = \text{exponential}(1)$, then $Y = X/\lambda$ is exponential(λ).

20. Suppose X is uniform on $(0, 1)$. Find the density function of $Y = X^n$.

21. Suppose X has density x^{-2} for $x \ge 1$ and $Y = X^{-2}$. Find the density function of Y.

22. Suppose X has an exponential distribution with parameter λ and $Y = X^{1/\alpha}$. Find the density function of Y. This is the *Weibull distribution.*

23. Suppose X has an exponential distribution with parameter 1 and $Y = \ln(X)$. Find the distribution function of X. This is the *double exponential distribution.*

24. Suppose X is uniform on $(0, \pi/2)$ and $Y = \sin X$. Find the density function of Y. The answer is called the *arcsine law* because the distribution function contains the arcsine function.

25. Suppose X has density function $f(x)$ for $-1 \le x \le 1$ and 0 otherwise. Find the density function of (a) $Y = |X|$ and (b) $Z = X^2$.

26. Suppose X has density function $x/2$ for $0 < x < 2$ and 0 otherwise. Find the density function of $Y = X(2 - X)$ by computing $P(Y \ge y)$ and then differentiating.

Joint distributions

27. Suppose X and Y have joint density $f(x, y) = c(x + y)$ for $0 < x, y < 1$. (a) What is c? (b) What is $P(X < 1/2)$?

28. Suppose X and Y have joint density $f(x, y) = 6xy^2$ for $0 < x, y < 1$. What is $P(X + Y < 1)$?

29. Suppose X and Y have joint density $f(x, y) = 2$ for $0 < y < x < 1$. Find $P(X - Y > z)$.

30. Suppose X and Y have joint density $f(x, y) = 1$ for $0 < x, y < 1$. Find $P(XY \leq z)$.

31. Two people agree to meet for a drink after work but they are impatient and each will wait only 15 minutes for the other person to show up. Suppose that they each arrive at independent random times uniformly distributed between 5 P.M. and 6 P.M. What is the probability they will meet?

32. Suppose X and Y have joint density $f(x, y) = e^{-(x+y)}$ for $x, y > 0$. Find the distribution function.

33. Suppose X is uniform on $(0, 1)$ and $Y = X$. Find the joint distribution function of X and Y.

34. A pair of random variables X and Y takes values between 0 and 1 and has $P(X \leq x, Y \leq y) = x^3y^2$ when $0 \leq x, y \leq 1$. Find the joint density function.

35. Given the joint distribution function $F_{X,Y}(x, y) = P(X \leq x, Y \leq y)$, how do you recover the marginal distribution function $F_X(x) = P(X \leq x)$?

36. Suppose X and Y have joint density $f(x, y)$. Are X and Y independent if

(a) $f(x, y) = xe^{-x(1+y)}$ for $x, y \geq 0$?
(b) $f(x, y) = 6xy^2$ when $x, y \geq 0$ and $x + y \leq 1$?
(c) $f(x, y) = 2xy + x$ when $0 < x < 1$ and $0 < y < 1$?
(d) $f(x, y) = (x + y)^2 - (x - y)^2$ when $0 < x < 1$ and $0 < y < 1$?

In each case, $f(x, y) = 0$ otherwise.

37. Suppose a point (X, Y) is chosen at random from the disk $x^2 + y^2 \leq 1$. Find (a) the marginal density of X and (b) the conditional density of Y given $X = x$.

38. Suppose X and Y have joint density $f(x, y) = x + 2y^3$ when $0 < x < 1$ and $0 < y < 1$. (a) Find the marginal densities of X and Y. (b) Are X and Y independent?

39. Suppose X and Y have joint density $f(x, y) = 6y$ when $x > 0$, $y > 0$, and $x + y < 1$. (a) Find the marginal densities of X and Y and (b) the conditional density of X given $Y = y$.

40. Suppose X and Y have joint density $f(x, y) = 10x^2 y$ when $0 < y < x < 1$. (a) Find the marginal densities of X and Y and (b) the conditional density of Y given $X = x$.

6

Limit Theorems

6.1 Sums of independent random variables

If X and Y are independent then

$$\begin{aligned} P(X+Y=z) &= \sum_x P(X=x, Y=z-x) \\ &= \sum_x P(X=x)P(Y=z-x) \end{aligned} \tag{6.1}$$

To see the first equality, note that if the sum is z then X must take on some value x and Y must be $z-x$. The first equality is valid for any random variables. The second holds since we have supposed X and Y are independent.

Example 6.1 Suppose X and Y are independent and have the following distribution:

k	1	2	3	4
Probability	0.1	0.2	0.3	0.4

Find the distribution of $X+Y$.

To compute a single probability is straightforward:

$$\begin{aligned} P(X+Y=4) &= \sum_{k=1}^{3} P(X=k)P(Y=3-k) \\ &= 0.1(0.3)+0.2(0.2)+0.3(0.1)=0.10 \end{aligned}$$

To compute the entire distribution, we begin by writing out the joint distribution of X and Y:

X	$Y=1$	2	3	4
1	0.01	0.02	0.03	0.04
2	0.02	0.04	0.06	0.08
3	0.03	0.06	0.09	0.12
4	0.04	0.08	0.12	0.16

The numbers with a given sum are diagonals in the table, so

$$P(X+Y=2) = 0.01$$
$$P(X+Y=3) = 0.02 + 0.02 = 0.04$$
$$P(X+Y=4) = 0.03 + 0.04 + 0.03 = 0.10$$
$$P(X+Y=5) = 0.04 + 0.06 + 0.06 + 0.04 = 0.20$$
$$P(X+Y=6) = 0.08 + 0.09 + 0.08 = 0.25$$
$$P(X+Y=7) = 0.12 + 0.12 = 0.24$$
$$P(X+Y=8) = 0.16$$

Example 6.2 If $X = \text{binomial}(n, p)$ and $Y = \text{binomial}(m, p)$ are independent then $X + Y = \text{binomial}(n + m, p)$.

Proof by thinking. The easiest way to see the conclusion is to note that if X is the number of successes in the first n trials and Y is the number of successes in the next m trials then $X + Y$ is the number of successes in $n + m$ trials.

Proof by computation. Using (6.1), noting that $P(X = j) = 0$ when $j < 0$ and $P(Y = k - j) = 0$ when $j > k$, and plugging in the definition of the binomial distribution we get

$$\begin{aligned} P(X+Y=k) &= \sum_{j=0}^{k} P(X=j)P(Y=k-j) \\ &= \sum_{j=0}^{k} C_{n,j}\, p^j(1-p)^{n-j} C_{m,k-j}\, p^{k-j}(1-p)^{m-(k-j)} \\ &= p^k(1-p)^{n+m-k} \sum_{j=0}^{k} C_{n,j} C_{m,k-j} \\ &= p^k(1-p)^{n+m-k} C_{n+m,k} \end{aligned}$$

To see the last equality, note that we can pick k students out of a class of n boys and m girls in $C_{n+m,k}$ ways, but this can also be done by first deciding on the number j of boys to be chosen and then picking j of the n boys (which can be done in $C_{n,j}$ ways) and $k - j$ of the m girls (which can be done in $C_{m,k-j}$ ways). The multiplication rule implies that for fixed j, the number of ways the

j boys and $k - j$ girls can be selected is $C_{n,j} C_{m,k-j}$, so summing from $j = 0$ to k gives

$$\sum_{j=0}^{k} C_{n,j} C_{m,k-j} = C_{n+m,k} \qquad \square$$

Since Poissons arise as the limit of binomials, we should guess that

Example 6.3 If $X = \text{Poisson}(\lambda)$ and $Y = \text{Poisson}(\mu)$ are independent then $X + Y = \text{Poisson}(\lambda + \mu)$.

Proof by thinking. Let $[x]$ be the largest integer $\leq x$, that is, x rounded down to the next integer. By Example 6.2 if $X = \text{Binomial}([n\lambda], 1/n)$ and $Y = \text{Binomial}([n\mu], 1/n)$ are independent, $X + Y = \text{Binomial}([n\lambda] + [n\mu], 1/n)$. Letting $n \to \infty$ and using the Poisson approximation to the binomial now gives the result.

Proof by computation. Again we use (6.1); note that $P(X = j) = 0$ when $j < 0$ and $P(Y = k - j) = 0$ when $j > k$, and plug in the definition of the Poisson distribution to get

$$\begin{aligned} P(X + Y = k) &= \sum_{j=0}^{k} P(X = j) P(Y = k - j) \\ &= \sum_{j=0}^{k} e^{-\lambda} \frac{\lambda^j}{j!} e^{-\mu} \frac{\mu^{k-j}}{(k-j)!} \\ &= e^{-(\lambda+\mu)} \frac{1}{k!} \sum_{j=0}^{k} C_{k,j} \lambda^j \mu^{k-j} \\ &= e^{-(\lambda+\mu)} \frac{(\lambda+\mu)^k}{k!} \end{aligned}$$

where the last equality follows from the binomial theorem (2.7).

Example 6.4 Suppose $X_1, \ldots, X_n$ are independent and have a geometric distribution with parameter p and let $T = X_1 + \cdots + X_n$ be the amount of time we have to wait for n successes when each trial is independent and results in success with probability p. Then

$$P(T = m) = C_{m-1,n-1} p^n (1 - p)^{m-n}$$

To see this consider the following possibility with $m = 10$ and $n = 3$:

FSFFFFSFFS

We know that there must be $m - n = 10 - 3 = 7$ failures and the last trial must be a success. In the first $m - 1 = 9$ trials, there must be $n - 1 = 2$ successes. Once we choose their locations, which can be done in $C_{m-1,n-1}$ ways, the outcome is determined. Each of these outcomes has probability $p^n(1-p)^{m-n}$.

6.1.1 Continuous random variables

Reasoning as in (6.1) the density function for the sum of two independent random variables with continuous distributions satisfies

$$f_{X+Y}(z) = \int f_X(x)\, f_Y(z-x)\, dx \tag{6.2}$$

Example 6.5 Suppose X and Y are independent and uniform on $(0, 1)$, then

$$f_{X+Y}(z) = \begin{cases} z & 0 \le z \le 1 \\ 2-z & 1 \le z \le 2 \end{cases}$$

Since the densities of X and Y are symmetric about 1/2, the density of the sum is symmetric about 1, and it suffices to prove the first half of the formula. Since $f_X(x)$ and $f_Y(y)$ are 1 on $[0, 1]$, if $0 \le z \le 1$,

$$f_{X+Y}(z) = \int_0^z 1\, dx = z$$

To generalize let f_k be the density function of the sum of k independent uniforms. By (6.2)

$$f_k(z) = \int_{z-1}^{z} f_{k-1}(y)\, dy$$

When $k = 3$ and $0 \le z \le 1$,

$$f_3(z) = \int_0^z y\, dy = \frac{z^2}{2}$$

When $2 \le z \le 3$, changing variables $x = 2 - y$,

$$f_3(z) = \int_{z-1}^{2} (2-y)\, dy = \int_0^{3-z} x\, dx = \frac{(3-z)^2}{2}$$

Since we expect the density to be symmetric about 3/2, it is comforting to note that $f_3(z) = f_3(3-z)$ for $2 \le z \le 3$.

The formula for middle piece $1 \le z \le 2$ takes more work

$$f_3(z) = \int_{z-1}^{1} y\,dy + \int_{1}^{z} (2-y)\,dy$$
$$= \frac{1}{2} - \frac{(z-1)^2}{2} + \left(2z - \frac{z^2}{2}\right) - \left(2 - \frac{1}{2}\right)$$

Using $-(z-1)^2 = -z^2 + 2z - 1$ with a little algebra gives

$$f_3(z) = \frac{-2z^2 + 6z - 3}{2} = z(3-z) - \frac{3}{2}$$

This answer has the expected symmetry and is $= 1/2 = z^2/2$ when $z = 1$.

It is remarkable that there is a general formula:

$$f_k(x) = \frac{1}{(k-1)!} \sum_{j=0}^{k-1} (-1)^j C_{k,j} (x-j)_+^{k-1} \tag{6.3}$$

where $y_+ = \max\{y, 0\}$ is the positive part of y. For a proof see Feller's *An Introduction to Probability Theory and Its Applications*, Vol. 2, pp. 27–28. To check the formula, note that when $k = 3$ and $0 \le x \le 1$, only the $j = 0$ term is not zero, so

$$f_3(x) = \frac{1}{2}x^2$$

When $1 \le x \le 2$, the $j = 1$ term is also nonzero and we get

$$f_3(x) = \frac{1}{2}[x^2 - 3(x-1)^2] = \frac{-2x^2 + 6x - 3}{2}$$

When $2 \le x \le 3$, we also have the $j = 2$ term, so

$$f_3(x) = \frac{1}{2}[-2x^2 + 6x - 3 + 3(x-2)^2] = \frac{1}{2}[x^2 - 6x + 9] = \frac{(3-x)^2}{2}$$

When $k = 4$, (6.3) gives

$$f_4(x) = \begin{cases} x^3/6 & 0 \le x \le 1 \\ \frac{1}{6}[x^3 - 3(x-1)^3] & 1 \le x \le 2 \end{cases}$$

and rest of the density can be computed from $f_4(x) = f_4(4-x)$ for $2 \le x \le 4$. The next figure graphs f_2, f_3, and f_4.

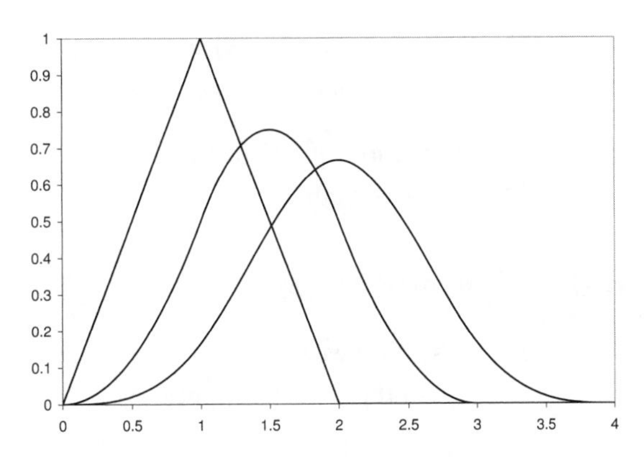

Example 6.6

X is said to have a gamma distribution with parameters n and λ or gamma(n, λ) for short, if

$$f_X(x) = \frac{\lambda^n x^{n-1}}{(n-1)!} e^{-x} \quad \text{for } x \geq 0$$

When $n = 1$, this reduces to the exponential(λ).

Theorem 6.1. *If $X = gamma(n, \lambda)$ and $Y = exponential(\lambda)$ then $X + Y = gamma(n, \lambda)$.*

Proof. Using (6.2) we have

$$f_{X+Y}(z) = \int_0^z \frac{\lambda^n x^{n-1}}{(n-1)!} e^{-x} \cdot \lambda e^{-(z-x)} \, dx$$

$$= \lambda^{n+1} e^{-z} \int_0^z \frac{x^{n-1}}{(n-1)!} \, dx = \frac{\lambda^{n+1} z^n}{n!} e^{-z} \qquad \square$$

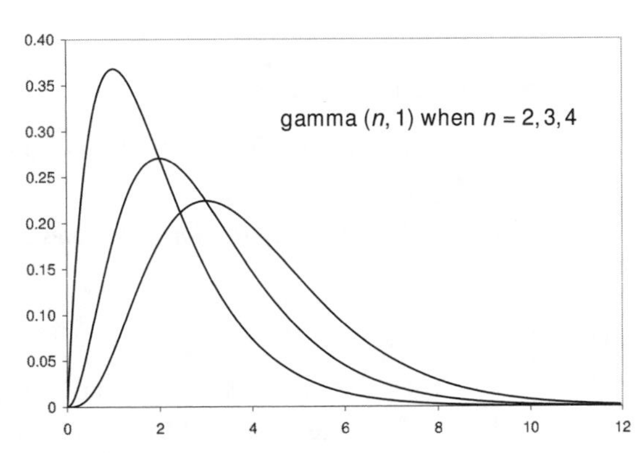

It follows from Theorem 6.1 that

(i) If $Y_1, \ldots, Y_n$ are independent exponential(λ), then the sum $Y_1 + \cdots + Y_n$ is gamma(n, λ).

(ii) If $X = \text{gamma}(m, \lambda)$ and $X = \text{gamma}(n, \lambda)$ are independent then $X + Y = \text{gamma}(m + n, \lambda)$.

The derivation of (ii) from (i) is the same as the "proof by thinking" for the binomial distribution.

6.2 Mean and variance of sums

In this section we show that the expected value of a sum of random variables is the sum of the expected values. We begin with the case of two random variables.

Theorem 6.2. *For any random variables X and Y,*

$$E(X + Y) = EX + EY \tag{6.4}$$

Proof. We have not derived a formula for the expected value of a function of two random variables, but the recipe is the same: sum the function times the probability over all possible pairs:

$$\begin{aligned} E(X + Y) &= \sum_{x,y}(x + y)\, P(X = x, Y = y) \\ &= \sum_{x,y} xP(X = x, Y = y) + \sum_{x,y} yP(X = x, Y = y) \end{aligned}$$

By the definition of the marginal distribution, $\sum_y P(X = x, Y = y) = P(X = x)$ and $\sum_x P(X = x, Y = y) = P(Y = y)$. This converts the above into

$$= \sum_x xP(X = x) + \sum_y yP(Y = y) = EX + EY$$

which is the desired result. □

From (6.4) and induction it follows that

Theorem 6.3. *For any random variables $X_1, \dots, X_n$,*

$$E(X_1 + \cdots + X_n) = EX_1 + \cdots + EX_n \tag{6.5}$$

Proof. (6.4) gives the result for $n = 2$. Applying (6.4) to $X = X_1 + \cdots + X_n$ and $Y = X_{n+1}$, we see that if the result is true for n, then

$$\begin{aligned} E(X_1 + \cdots + X_{n+1}) &= E(X_1 + \cdots + X_n) + EX_{n+1} \\ &= EX_1 + \cdots + EX_n + EX_{n+1} \end{aligned} \tag{6.6}$$

and the result holds for $n + 1$. □

Formula (6.5) is very useful in doing computations.

Example 6.7

Pick two cards out of a deck of 52 and let X be the number of spades. Calculate the expected value of X.

To do this directly from the definition we have to calculate the distribution

$$P(X=2) = \frac{C_{13,2}}{C_{52,2}} = \frac{13 \cdot 12}{52 \cdot 51}$$

$$P(X=1) = \frac{C_{13,1}C_{39,1}}{C_{52,2}} = \frac{2 \cdot 13 \cdot 39}{52 \cdot 51}$$

From this it follows that

$$EX = 2 \cdot \frac{13 \cdot 12}{52 \cdot 51} + \frac{2 \cdot 13 \cdot 39}{52 \cdot 51} = 2 \cdot \frac{13 \cdot 51}{52 \cdot 51} = \frac{1}{2}$$

To see this more easily, let $X_i = 1$ if the ith card drawn was a spade and 0 otherwise. $X = X_1 + X_2$, so it follows from (6.5) that

$$EX = EX_1 + EX_2 = 1/4 + 1/4 = 1/2$$

since $P(X_i = 1) = 1/4$ and $P(X_i = 0) = 3/4$.

In a similar way we can conclude that if we draw 13 cards out of a deck of 52, the expected number of spades is $13/4 = 3.25$. Doing this directly from the distribution would be extremely tedious. In our next example it would be difficult to calculate the distribution, but it is easy to compute the expected value.

Example 6.8

Balls in boxes. Suppose we put n balls randomly into m boxes. What is the expected number of empty boxes?

Let $X_i = 1$ if the ith box is empty. The total number of empty boxes is given by $N = X_1 + \cdots + X_m$, so it follows from (6.5) that

$$EN = EX_1 + \cdots + EX_m = mEX_1$$

The probability that box 1 is empty is $(1 - 1/m)^n$, so the expected number of empty boxes is $m(1 - 1/m)^n$.

For a concrete example, suppose $m = 100$ and $n = 500$. In this case,

$$(1 - 1/100)^{500} \approx e^{-5} = 0.00673$$

so the expected number of empty boxes is 2/3 (approx.). We will return to this computation in Example 6.17.

The next goal of this section is to show that if $X_1, \ldots, X_n$ are independent then

$$\text{var}\,(X_1 + \cdots + X_n) = \text{var}\,(X_1) + \cdots + \text{var}\,(X_n).$$

The first step is to prove

Theorem 6.4. *If X and Y are independent then*

$$EXY = EX \cdot EY \tag{6.7}$$

Proof. Since X and Y are independent, $P(X = x, Y = y) = P(X = x)P(Y = y)$ and

$$EXY = \sum_{x,y} xyP(X = x)P(Y = y) = \sum_y yP(Y = y) \sum_x xP(X = x)$$
$$= \sum_y yP(Y = y)EX = EX \cdot EY$$

which proves the desired result. □

(6.4) says that $E(X + Y) = EX + EY$ holds for ANY random variables. The next example shows that $EXY = EX \cdot EY$ does not hold in general.

Example 6.9

Suppose X and Y have joint distribution given by

X	$Y = 1$	0
1	0	0.3
0	0.5	0.2

We have arranged things so that XY is always 0, so $EXY = 0$. On the other hand, $EX = P(X = 1) = 0.3$ and $EY = P(Y = 1) = 0.5$, so

$$EXY = 0 < 0.15 = EX\,EY$$

Our next example shows that we may have $EXY = EX\,EY$ without X and Y being independent.

Example 6.10

Suppose X and Y have joint distribution given by

X	$Y = -1$	0	1	
1	0	0.25	0	0.25
0	0.25	0.25	0.25	0.75
	0.25	0.5	0.25	

Again we have arranged things so that XY is always 0, so $EXY = 0$. The symmetry of the marginal distribution for Y (or simple arithmetic) shows $EY = 0$, so we have $EXY = 0 = EX\,EY$. X and Y are not independent since

$$P(X = 1, Y = -1) = 0 < P(X = 1)P(Y = -1).$$

Our next topic is the variance of sums. To state our first result we need a definition. The **covariance** of X and Y is

$$\text{cov}(X, Y) = E\{(X - EX)(Y - EY)\}$$

Now $\text{cov}(X, X) = E\{(X - EX)^2\} = \text{var}(X)$, so repeating the proof of (1.11) we can rewrite the definition of the covariance in a form that is more convenient for computations:

$$\text{cov}(X, Y) = E\{XY - YEX - XEY + EXEY\} = EXY - EXEY \quad (6.8)$$

Remembering (6.7), we see that if X and Y are independent, $\text{cov}(X, Y) = 0$.

Example 6.11 Consider the calculus grade joint distribution (Example 3.27)

X	$Y = 4$	3	2	
5	0.10	0.05	0	0.15
4	0.15	0.15	0	0.30
3	0.10	0.15	0.10	0.35
2	0	0.05	0.10	0.15
1	0	0	0.05	0.05
	0.35	0.40	0.25	

$EX = 5(0.15) + 4(0.30) + 3(0.35) + 2(0.15) + 1(0.05) = 3.35$ and $EY = 4(0.35) + 3(0.40) + 2(0.25) = 3.10$. Patiently adding up all the possibilities we see $E(XY) = 10.9$, so

$$\text{cov}(X, Y) = 10.9 - (3.35)(3.10) = 0.515 > 0$$

Positive covariance means that when one variable is above average then the other has a greater tendency to be above average. For example, we expect a person's height and weight to have this relationship. In the opposite direction is

Example 6.12 The urn joint distribution (Example 3.26)

X	$Y = 0$	1	2	
0	6/105	20/105	10/105	36/105
1	24/105	30/105	0	54/104
2	15/105	0	0	15/105
	45/105	50/105	10/105	

$$EX = (54/105) \cdot 1 + (15/105) \cdot 2 = 84/105 = 4/5$$

$$EY = (50/105) \cdot 1 + (10/105) \cdot 2 = 70/105 = 2/3$$

$$EXY = (30/105) \cdot 1 = 2/7$$

so we have

$$\text{cov}(X, Y) = \frac{2}{7} - \frac{4}{5} \cdot \frac{2}{3} = -0.2476$$

Negative covariance means that when one variable is above average then the other has a greater tendency to be below average. For example, we expect for a high school student, the number of hours of TV watched and their grades to have this relationship. In the urn example, negative covariance is clear since there is a triangle of 0's in the joint distribution that comes from the restriction $X + Y \leq 2$.

The next result explains our interest in the covariance

$$\text{var}(X + Y) = \text{var}(X) + 2\,\text{cov}(X, Y) + \text{var}(Y) \tag{6.9}$$

Proof. Using $E(X + Y) = EX + EY$ and working out the square

$$\begin{aligned} E(X + Y - E(X + Y))^2 &= E((X - EX) + (Y - EY))^2 \\ &= E(X - EX)^2 + 2E((X - EX)(Y - EY)) \\ &\quad + E(Y - EY)^2 \\ &= \text{var}(X) + 2\,\text{cov}(X, Y) + \text{var}(Y) \end{aligned}$$

which proves the desired formula. □

Reasoning similar to the proof of (6.9) with more algebra leads to

$$\text{var}\left(\sum_{i=1}^{n} X_i\right) = \sum_{i=1}^{n} \text{var}(X_i) + 2 \sum_{1 \leq i < j \leq n} \text{cov}(X_i, X_j) \tag{6.10}$$

The most important special case of (6.10) is

Theorem 6.5. *If $X_1, \ldots, X_n$ are independent then*

$$var(X_1 + \cdots + X_n) = var(X_1) + \cdots + var(X_n). \tag{6.11}$$

Proof. Our assumption implies that the second sum in (6.10) vanishes. □

(6.11) is useful in computing variances. For example, it reduces the computation of the variance of the binomial to the following trivial case.

Example 6.13

Bernoulli distribution. $X = 1$ with probability p and 0 with probability $1 - p$. $EX = 1 \cdot p + 0 \cdot (1 - p) = p$ and $E(X^2) = 1^2 \cdot p + 0^2 \cdot (1 - p) = p$, so $\text{var}(X) = p - p^2 = p(1 - p)$.

Example 6.14

Binomial distribution. Consider a sequence of independent trials in which success has probability p. Let $X_i = 1$ if the ith trial results in a success and 0 otherwise. The total number of successes, $S_n = X_1 + \cdots + X_n$, has a binomial(n, p) distribution. Compute the variance of X.

(6.11) and Example 6.13 imply

$$\text{var}\,(S_n) = n \,\text{var}\,(X_1) = np(1-p).$$

To see what this says consider $n = 400$ and $p = 1/2$; that is, S_n is the number of heads when we flip 400 coins. (6.5) implies that $E\,S_{400} = 400(1/2)$, so the expected number of heads is 200. $\text{var}\,(S_{400}) = 400(1/2)(1/2) = 100$, so the standard deviation $\sigma(S_{400})$ is 10.

Informally, in 400 tosses we expect 200 ± 10 heads, where $\pm$ can be read as "give or take." In Section 6.5, we make this more precise: the number of heads will be in $[190, 210]$ 68% of the time and in $[180, 220]$ 95% of the time.

Example 6.15

Hypergeometric distribution. Suppose we have an urn with n balls, m of which are red. Let R be the number of red balls we get when we draw from the urn k times without replacement. If we let $X_i = 1$ if the ith draw is red, and 0 otherwise, then $R = X_1 + \cdots + X_k$. From this we see that

$$E\,R = kE\,X_1 = kp \quad \text{where } p = m/n.$$

To compute the variance we will use (6.10). Since the X_i are Bernoulli random variables, $\text{var}\,(X_i) = p(1-p)$. As for the covariance,

$$E(X_1X_2) = P(X_1 = 1, X_2 = 1) = \frac{m}{n} \cdot \frac{m-1}{n-1}$$

so we have

$$\begin{aligned}\text{cov}\,(X_1, X_2) &= \frac{m}{n} \cdot \frac{m-1}{n-1} - \frac{m}{n} \cdot \frac{m-1}{n-1} \\ &= \frac{m}{n} \cdot \left(\frac{(m-1)n - m(n-1)}{n(n-1)} \right) = -\frac{m}{n} \cdot \frac{n-m}{n(n-1)}\end{aligned}$$

and it follows from (6.10) that

$$\text{var}\,(R) = kp(1-p) - k(k-1)p(1-p)\frac{1}{n-1} = kp(1-p)\left[1 - \frac{k-1}{n-1}\right]$$

It is comforting to note that $\text{var}\,(R) = 0$ when $k = n$, that is, when we draw all of the balls out of the urn. If $k = xn$, then $\text{var}\,(R) \approx nx(1-x) \cdot p(1-p)$, so the maximum occurs when $k \approx n/2$.

What have we learned?

The most important facts from this section are

(6.5) *For any random variables* $X_1, \ldots, X_n$,

$$E(X_1 + \cdots + X_n) = EX_1 + \cdots + EX_n$$

(6.11) *If* $X_1, \ldots, X_n$ *are independent then*

$$\text{var}(X_1 + \cdots + X_n) = \text{var}(X_1) + \cdots + \text{var}(X_n)$$

These two facts imply that if S_n is a sum of n independent random variables with mean μ and variance σ^2, then

$$ES_n = n\mu \quad \text{var}(S_n) = n\sigma^2$$

and hence the size of the typical deviation from the mean is $\sigma\sqrt{n}$.

To illustrate the use of these results, we consider two more examples.

Example 6.16

Rolling dice. Suppose we roll two dice 100 times. What is the mean, variance, and standard deviation of the sum of the points?

This is equivalent to rolling one die 200 times. Let S_{200} be the sum. (6.5) implies that $ES_{200} = 200(7/2) = 700$, while (6.11) and Example 1.26 imply $\text{var}(S_{200}) = 200(105/36) = 583.33$, so $\sigma(S_{200}) = 24.15$. Informally, the sum will be 700 give or take 24.

In our final example the variables being summed do not all have the same distribution.

Example 6.17

Coupon collector's problem. Suppose that we record the birthday of every person we meet. Let N be the number of people we have to meet until we have seen someone with every birthday. Ignoring February 29, find the mean and variance of N.

Let T_k be the time at which we see our kth different birthday, so $N = T_{365}$. If we let $T_0 = 0$, then for $0 \le k \le 364$, $T_{k+1} - T_k$ is geometric with success probability $1 - k/365$, since up to that point we have collected k birthdays. Recalling that the mean of the geometric with success probability p is $1/p$, this is $365/(365 - k)$, and summing gives

$$ET_{365} = \sum_{k=0}^{364} \frac{365}{365 - k} = 365 \sum_{j=1}^{365} \frac{1}{j} = 2{,}364.64$$

Since the variance of geometric(p) is $(1-p)/p^2$,

$$\text{var}(T_{365}) = \sum_{k=1}^{364} \frac{k/365}{(1-k/365)^2} = \sum_{j=1}^{364} \frac{1-j/365}{(j/365)^2}$$

$$= (365)^2 \sum_{j=1}^{364} \frac{1}{j^2} - 365 \sum_{k=1}^{364} \frac{1}{j} = 216,418$$

The variance looks frighteningly large until you take square root to conclude that the standard deviation is 465.2. Informally, we will need to meet 1,900–2,930 people to see all 365 birthdays.

General formulas. Suppose now that there are n objects to be collected. Let T_k be the time at which we see our kth different object, so $N = T_n$. $T_1 = 1$. For $1 \le k \le n-1$, $T_{k+1} - T_k$ is geometric with success probability $1 - k/n$, since up to that point we have collected k objects. Using our formula for the mean of the geometric,

$$E\,T_n = n \sum_{j=1}^{n-1} \frac{1}{j} \approx n \ln n$$

since $\sum_{i=1}^{n-1} 1/j \approx \int_1^n dx/x = \ln n$. Turning to the variance,

$$\text{var}(T_n) = \sum_{j=1}^{n-1} \frac{1-j/n}{(j/n)^2}$$

$$= n^2 \sum_{j=1}^{n-1} \frac{1}{j^2} - n \sum_{j=1}^{n-1} \frac{1}{j^2} \approx n^2\pi^2/6$$

since $\sum_{j=1}^{\infty} 1/j^2 = \pi^2/6$. Note that the mean is mean $n \ln n$, while the standard deviation is of order n.

Back to balls in boxes, Example 6.8. When $n = 100$, $n \ln n = 460.51$ and the standard deviation is roughly $n\pi/\sqrt{6} = 128.25$. This seems consistent with the previous computation that putting 500 balls into 100 boxes leaves an average of 2/3 empty boxes.

6.3 Laws of large numbers

To motivate the main result of this section, we turn to simulation. The next figure shows the fraction of heads versus time in a simulation of flipping a fair coin 5,000 times. As dictated by the frequency interpretation of probability, the fraction of heads seen at time n approaches 1/2 as $n \to \infty$.

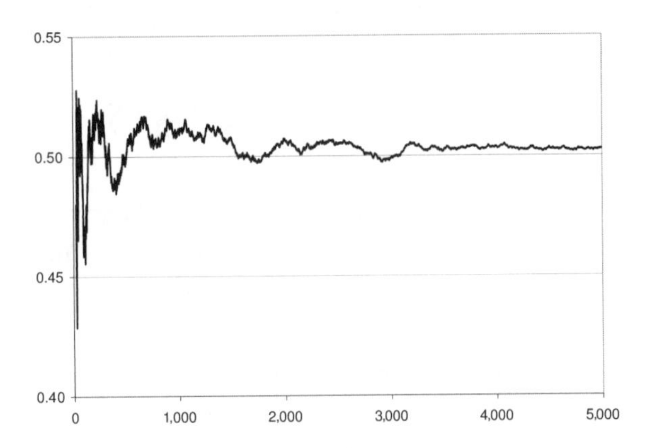

To state the general result, we need some definitions. If $X_1, X_2, \ldots,$ are independent and have the same distribution then we say the X_i are **independent and identically distributed**, or i.i.d. for short. Such sequences arise if we repeat some experiment, such as flipping a coin or rolling a die, or if we stop people at random and measure their height or ask them how they will vote in an upcoming election.

Our first goal in this section is to prove the **law of large numbers**, which says that if $X_1, X_2, \ldots,$ are i.i.d. with $EX_i = \mu$, then when n is large, the average of the first n observations,

$$\bar{X}_n = (X_1 + \cdots + X_n)/n$$

is close to EX with high probability.

$\bar{X}_n$ is called the **sample mean** because if we assigned probability $1/n$ to each of the first n observations then $\bar{X}_n$ would be the mean of that distribution. If we suppose that the X_i are i.i.d. with $EX_i = \mu$, then using the facts that $E(cY) = cEY$ and the expected value of the sum is the sum of the expected values, we have

$$\begin{aligned} E\bar{X}_n &= \frac{1}{n}E(X_1 + \cdots + X_n) \\ &= \frac{1}{n}(EX_1 + \cdots + EX_n) = \mu \end{aligned} \tag{6.12}$$

If we suppose that $\text{var}(X_i) = \sigma^2$, then using the facts that $\text{var}(cY) = c^2\,\text{var}(Y)$ and that for independent $X_1, \ldots, X_n$ the variance of the sum is the sum of the variances, we have

$$\text{var}(\bar{X}_n) = \frac{1}{n^2}\text{var}(X_1 + \cdots + X_n)$$
$$= \frac{1}{n^2}(\text{var}(X_1) + \cdots + \text{var}(X_n)) = \frac{\sigma^2}{n} \quad (6.13)$$

Taking square roots we see that the standard deviation of $\bar{X}_n$ is $\sigma/\sqrt{n}$. We have earlier called this the size of a typical deviation from the mean. The key to proving the law of large numbers is to show that if k is large then the probability of an observation more than k standard deviations from the mean is small. To motivate the inequality we will use to prove this, we consider a

Puzzle. Suppose $EX = 0$ and $EX^2 = 1$. How large can $P(|X| \geq 3)$ be?

Solution. On the set $\{|X| \geq 3\}$, $X^2 \geq 9$. Since $X^2 \geq 0$, EX^2 must be larger than what we get from considering only values with $|X| \geq 3$. That is,

$$1 = EX^2 \geq 9P(|X| \geq 3)$$

or $P(|X| \geq 3) \leq 1/9$. To see that this can be achieved, we let $P(X = 3) = 1/18$, $P(X = -3) = 1/18$, and $P(X = 0) = 8/9$ and note that

$$EX = 3 \cdot \frac{1}{18} + (-3) \cdot \frac{1}{18} = 0$$
$$EX^2 = 9 \cdot \frac{1}{18} + 9 \cdot \frac{1}{18} = 1$$

Generalizing leads to

Chebyshev's inequality. *If $y > 0$, then*

$$P(|Y - EY| \geq y) \leq \text{var}(Y)/y^2 \quad (6.14)$$

Proof. Again since $|Y - EY|^2 \geq 0$, $E|Y - EY|^2$ must be larger than what we get from considering only values with $|Y - EY| \geq y$ so

$$\text{var}(Y) = E|Y - EY|^2 \geq y^2 P(|Y - EY| \geq y)$$

and rearranging gives (6.14). □

If we let $\sigma^2 = \text{var}(Y)$ and take $y = k\sigma$ with $k \geq 1$, then (6.14) implies that

$$P(|Y - EY| \geq k\sigma) \leq 1/k^2 \quad (6.15)$$

This reinforces our notion that σ is the size of the typical deviation from the mean by showing that a deviation of k standard deviations has probability smaller than $1/k^2$.

Proof of the law of large numbers. Let $Y = \bar{X}_n$. (6.12) implies $EY = \mu$ and (6.13) implies $\text{var}(Y) = \sigma^2/n$, so using Chebyshev's inequality, we see that if $\epsilon > 0$, then

$$P(|\bar{X}_n - \mu| \geq \epsilon) \leq \frac{\text{var}(\bar{X}_n)}{\epsilon^2} = \frac{\sigma^2}{\epsilon^2 n} \to 0 \tag{6.16}$$

as $n \to \infty$. □

(6.16) could be called the **fundamental theorem of statistics** because it says that the sample mean is close to the mean μ of the underlying population when the sample is large. The last conclusion does not rule out the possibility that the sequence of sample means $\bar{X}_1, \bar{X}_2, \ldots,$ stays close to EX most of the time but occasionally wanders off because of a streak of bad luck. Our next result says that this does not happen.

Strong law of large numbers. *Suppose $X_1, X_2, \ldots,$ are i.i.d. with $E|X_i| < \infty$. Then with probability 1 the sequence of numbers $\bar{X}_n$ converges to EX_i as $n \to \infty$.*

The first thing we have to explain is the phrase "with probability 1." To do this we first consider flipping a coin and letting X_i be 1 if the ith toss results in heads and 0 otherwise. The strong law of large numbers says that with probability 1,

$$(X_1 + \cdots + X_n)/n \to 1/2 \quad \text{as } n \to \infty$$

It is easy to write down sequences of tosses for which this is false:

H, H, T, H, H, T, H, H, T, H, H, T, ...

However, the strong law of large numbers implies that the collection of "bad sequences" (that is, those for which the asymptotic frequency of heads is not 1/2) has probability zero.

Example 6.18 **Growth of a risky investment.** In our simple model, each year the stock market either increases by 40% or decreases by 20% with probability 1/2 each. How fast will our money grow?

Reasoning naively, our expected value is $(1/2)40 + (1/2)(-20) = 10\%$ per year. However, this is not the right way to look at things. Let X_i be independent and equal to 1.4 and 0.8 with probability 1/2 each. If we start with M_0 dollars then after n years we have

$$M_n = M_0 X_1 X_2 \cdots X_n$$

To turn this into something to which we can apply the law of large numbers, we take logarithms:

$$\ln(M_n/M_0) = \sum_{i=1}^{n} \ln X_i \approx nE(\ln X_i)$$

Computing the expected value,

$$E(\ln X_i) = \frac{1}{2}\ln(1.4) + \frac{1}{2}\ln(0.8) = 0.0566643$$

so we have

$$M_n \approx M_0 \exp(n(\ln(1.4) + \ln(0.8))/2) = M_0(\sqrt{(1.4)(0.8)})^n$$

From this we see that the growth is not given by the arithmetic mean $(1.4 + 0.8)/2 = 1.1$, but by the geometric mean $\sqrt{(1.4)(0.8)} = 1.0583$.

Example 6.19

Optimal investment. Suppose now that we can either (i) invest in the stock market with either increases by 40% or decreases by 20% with probability 1/2 each or (ii) buy a bond that always pays 4% interest. How should we allocate our money between these two alternatives to maximize our rate of return?

The answer is easier to understand if we consider a general problem. Let $\alpha = 1.4$ and $\beta = 0.8$ be the two outcomes for the stock and let $\gamma = 1.04$ be the growth for the bond. Let M_n be the amount of money we have at the end of year n. If we put a fraction p in the bond and $1 - p$ in the stock market then the expected return is

$$R(p) = E(\ln(M_1/M_0)) = \frac{1}{2}\ln[p\gamma + (1-p)\alpha] + \frac{1}{2}\ln[p\gamma + (1-p)\beta]$$

To optimize we take the derivative:

$$\begin{aligned} R'(p) &= \frac{1}{2}\frac{\gamma - \alpha}{\alpha + p(\gamma - \alpha)} + \frac{1}{2}\frac{\gamma - \beta}{\beta + p(\gamma - \beta)} \\ &= \frac{\beta(\gamma - \alpha) + \alpha(\gamma - \beta) + 2(\gamma - \beta)(\gamma - \alpha)p}{2(\alpha + p(\gamma - \alpha))(\beta + p(\gamma - \beta))} \end{aligned}$$

For the maximum to occur at a value $0 < p < 1$, we need $R'(0) > 0$ and $R'(1) < 0$. The denominator is positive. If we let $q(p)$ be the numerator of $R'(p)$,

$$\begin{aligned} q(0) &= \gamma(\beta + \alpha) - 2\alpha\beta \\ q(1) &= \gamma(\gamma - \alpha) + \gamma(\gamma - \beta) \end{aligned}$$

For $q(1) < 0$, that is, to have some of the stock in our portfolio, we need $\gamma < (\alpha + \beta)/2$. For $q(0) > 0$, that is, to have some of the bond in our portfolio, we need

$$\gamma > \frac{\alpha\beta}{(\alpha + \beta)/2}$$

When both these conditions hold there will be an optimal that can be found by setting $q(p^*) = 0$.

The next figure gives the return on investment as a function of p in our concrete example: $\alpha = 1.4$, $\beta = 0.8$, and $\gamma = 1.04$. To connect the value at 0 with Example 6.18, we note that the exponential growth rate is $\ln(1.0583) = 0.0566$.

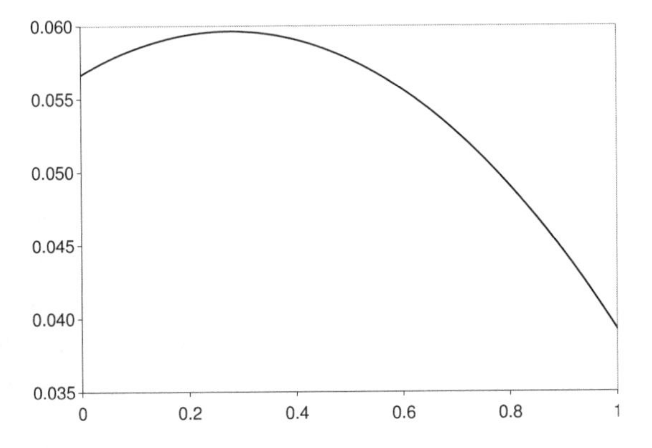

To find the maximum analytically,

$$\frac{\alpha + \beta}{2} = 1.1 > \gamma = 1.04 > \frac{\alpha\beta}{(\alpha + \beta)/2} = \frac{(1.4)(0.8)}{1.1} = 1.018$$

so the optimal fraction in the bond is

$$p^* = \frac{\gamma(\beta + \alpha)/2 - \alpha\beta}{(\gamma - \beta)(\alpha - \gamma)} = \frac{(1.04)(1.1) - (1.4)(0.8)}{(0.24)(0.36)} = 0.2777$$

6.4 Normal distribution

To prepare for results in the next section we need to introduce the **standard normal distribution**

$$f(x) = (2\pi)^{-1/2} e^{-x^2/2}$$

which looks like

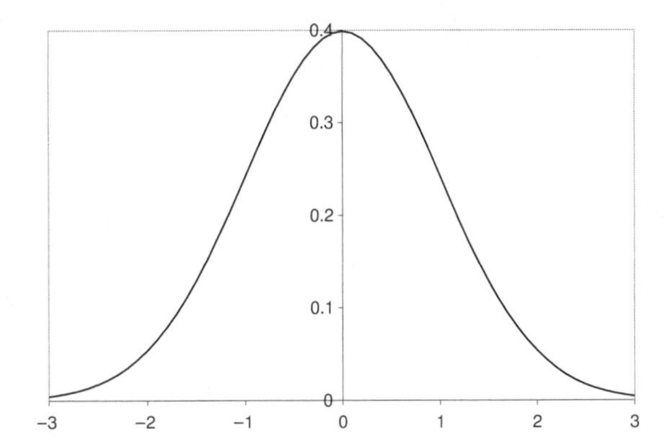

Since there is no closed-form expression for the antiderivative of f, it takes some ingenuity to check:

Theorem 6.6. *$f(x) = (2\pi)^{-1/2}e^{-x^2/2}$ is a probability density.*

Proof. Let $I = \int e^{-x^2/2}\,dx$. To show that $\int f(x)\,dx = 1$, we want to show that $I = \sqrt{2\pi}$.

$$I^2 = \int e^{-x^2/2}\,dx \int e^{-y^2/2}\,dy = \iint e^{-(x^2+y^2)/2}\,dx\,dy$$

Changing to polar coordinates, the last integral becomes

$$\int_0^\infty \int_0^{2\pi} e^{-r^2/2}\,r\,d\theta\,dr = 2\pi \int_0^\infty e^{-r^2/2}\,r\,dr = 2\pi\,(-e^{-r^2/2})\Big|_0^\infty = 2\pi$$

So $I^2 = 2\pi$ or $I = \sqrt{2\pi}$. □

There is, unfortunately, no closed-form expression for the distribution function,

$$\Phi(x) = \int_{-\infty}^{x} (2\pi)^{-1/2} e^{-y^2/2}\,dy$$

so we have to use a table like the one given in the Appendix. In the next few calculations, we will only use a little of the table:

x	0	1	2	3
$\Phi(x)$	0.5000	0.8413	0.9972	0.9986

$P(a < X \le b) = F(b) - F(a)$. For example, when $b = 2$ and $a = 1$, we have

$$P(1 < X \le 2) = \Phi(2) - \Phi(1) = 0.9772 - 0.8413 = 0.1359$$

Our little table and the one in the back of the book only give the values of $\Phi(x)$ for $x \geq 0$. Values for $x < 0$ are computed by noting that the normal density function is symmetric about 0 ($f(x) = f(-x)$), so

$$P(X \leq -x) = P(X \geq x) = 1 - P(X \leq x)$$

since $P(X = x) = 0$. For an example of the use of symmetry, we note that

$$P(X \leq -1) = 1 - P(X \leq 1) = 1 - 0.8413 = 0.1587$$

so we have

$$P(-1 \leq X \leq 1) = \Phi(1) - \Phi(-1) = 0.8413 - 0.1587 = 0.6826$$

In the case of discrete random variables, it was important to keep track of the difference between $<$ and $\leq$. Here it is not, since for the normal distribution $P(X = x) = 0$ for all x.

The standard normal distribution has mean 0 by symmetry:

$$\mathrm{var}\,(X) = E(X^2) = \int (2\pi)^{-1/2} x^2 e^{-x^2/2}\, dx$$

To show that the variance is 1, we use integration by parts (5.7), with $g(x) = (2\pi)^{-1/2}x$, $h'(x) = xe^{-x^2/2}$, and hence $g'(x) = (2\pi)^{-1/2}$, $h(x) = -e^{-x^2/2}$ to conclude that the integral above is

$$\begin{aligned} &= -(2\pi)^{-1/2} x e^{-x^2/2}\Big|_{-\infty}^{\infty} + \int (2\pi)^{-1/2} e^{-x^2/2}\, dx \\ &= 0 + 1 \end{aligned}$$

For the last step we observe that $e^{-x^2/2}$ goes to 0 much faster than x goes to ∞, and the integral gives the total mass of the normal density.

To create a normal distribution with mean μ and variance σ^2, let $Y = \sigma X + \mu$, where $\sigma > 0$. The inverse of $r(x) = \sigma x + \mu$ is $s(y) = (y - \mu)/\sigma$, so (5.9) implies that Y has density function

$$\begin{aligned} f(s(y))s'(y) &= (2\pi)^{-1/2} e^{-\{(y-\mu)/\sigma\}^2/2} \frac{1}{\sigma} \\ &= (2\pi\sigma^2)^{-1/2} e^{-(y-\mu)^2/2\sigma^2} \end{aligned}$$

If $Y = \text{normal}(\mu, \sigma^2)$, then reversing the formula we used to define Y, $X = (Y - \mu)/\sigma$ has the standard normal distribution.

Example 6.20

Suppose that a man's height has a normal distribution with mean $\mu = 69$ inches and standard deviation $\sigma = 3$ inches. What is the probability that a randomly chosen man is more than 6 feet tall (72 inches)?

The first step in the solution is to rephrase the question in terms of X:

$$P(Y \geq 72) = P(Y - 69 \geq 3) = P\left(\frac{Y-69}{\sqrt{9}} \geq 1\right)$$
$$= P(X \geq 1) = 1 - 0.8413 = 0.1587$$

from the little table above.

An important property of the normal is

Theorem 6.7. *If $X = normal(\mu, a)$ and $Y = normal(\nu, b)$ are independent then $X + Y = normal(\mu + \nu, a + b)$.*

Proof. By considering $X' = X - \mu$ and $Y' = Y - \nu$ and noting $X + Y = X' + Y' + (\mu + \nu)$, it suffices to prove the result when $\mu = \nu = 0$. By (6.2) we need to evaluate

$$\int \frac{1}{\sqrt{2\pi a}} e^{-z^2/2a} \frac{1}{\sqrt{2\pi b}} e^{-(x-z)^2/2b} \, dz$$

The product of the two exponentials is

$$\exp\left(-\frac{z^2}{2a} - \frac{z^2}{2b} + \frac{2xz}{2b} - \frac{x^2}{2b}\right)$$

The quantity in the exponent is

$$-\frac{z^2(a+b)}{2ab} + \frac{2axz}{2ab} - \frac{ax^2}{2ab}$$
$$= \frac{a+b}{2ab}\left(-z^2 + \frac{2a}{a+b}xz - \frac{a}{a+b}x^2\right)$$
$$= \frac{a+b}{2ab}\left[-\left(z - \frac{ax}{a+b}\right)^2 - \frac{ab}{(a+b)^2}x^2\right]$$

Working backward through our steps we need to evaluate

$$\frac{1}{\sqrt{2\pi a}}\frac{1}{\sqrt{2\pi b}} e^{-x^2/2(a+b)} \cdot \int \exp\left(-\frac{a+b}{2ab}\left(z - \frac{ax}{a+b}\right)^2\right) dz$$

The integrand is a constant multiple of the density of a normal with mean $ax/(a+b)$ and variance $ab/(a+b)$, so it is $= \sqrt{2\pi ab/(a+b)}$ and we end up with

$$\frac{1}{\sqrt{2\pi(a+b)}} e^{-x^2/2(a+b)}$$

which is the desired density function. □

6.5 Central limit theorem

The limit theorem in this section gets its name not only from the fact that it is of central importance but also because it shows that if you add up a large number of random variables with a fixed distribution with finite variance, then if we subtract the mean and divide by the standard deviation, the result has approximately a normal distribution.

Central limit theorem (CLT). *Suppose $X_1, X_2, \ldots,$ are i.i.d. and have $EX_i = \mu$ and $var(X_i) = \sigma^2$ with $0 < \sigma^2 < \infty$. Let $S_n = X_1 + \cdots + X_n$. As $n \to \infty$,*

$$P\left(a \le \frac{S_n - n\mu}{\sigma\sqrt{n}} \le b\right) \to \int_a^b \frac{1}{\sqrt{2\pi}} e^{-x^2/2}\, dx \tag{6.17}$$

To do computations it is convenient to introduce χ, a normal random variable with mean 0 and 1. With this we can write the last result as

$$\frac{S_n - n\mu}{\sigma\sqrt{n}} \approx \chi$$

Some people seem mystified that I use chi for the normal, but to me it seems natural since the square of a normal has a chi-squared distribution.

Example 6.21 Suppose we flip a coin 100 times. What is the probability that we get at least 56 heads?

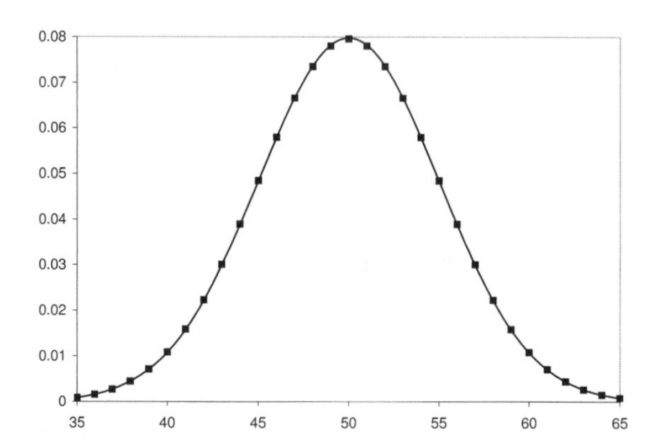

As the figure shows, the distribution (given by squares) is approximately normal (the smooth curve). To use (6.17) we note that $X_i = 1$ or 0 with probability 1/2 each, so the X_i have mean 1/2 and standard deviation $\sqrt{0.5 \cdot 0.5} = 1/2$. The mean number of heads in 100 tosses is $100/2 = 50$ and the standard deviation

$\sqrt{100}/2 = 5$, so (6.17) implies

$$P(S_{100} \geq 56) = P\left(\frac{S_{100} - 50}{5} \geq \frac{6}{5}\right) \approx P(\chi \geq 1.2)$$
$$= 1 - P(\chi \leq 1.2) = 1 - 0.8849 = 0.1151$$

As we will now explain there is a small problem with this solution. If the question in the problem had been formulated as "What is the probability of at most 55 heads?" we would have computed

$$P(S_{100} \leq 55) = P\left(\frac{S_{100} - 450}{5} \leq \frac{5}{5}\right) \approx P(\chi \leq 1.0) = 0.8413$$

which does not quite agree with our first answer since

$$0.8413 + 0.1151 = 0.9568 < 1$$

whereas $P(S_{100} \leq 55) + P(S_{100} \geq 56) = 1$. The solution to this problem is to regard $\{S_{100} \geq 56\}$ as $\{S_{100} \geq 55.5\}$; that is, the integers 55 and 56 split up the territory that lies between them:

55 56

When we do this, the answer to our original question becomes

$$P(S_{100} \geq 55.5) = P\left(\frac{S_{100} - 50}{55} \geq \frac{5.5}{15}\right)$$
$$\approx P(\chi \geq 1.1) = 1 - 0.8643 = 0.1357$$

which is a much better approximation of the exact probability 0.135627 than was our first answer, 0.1151.

The last correction, which is called the **histogram correction**, should be used whenever we apply (6.17) to integer-valued random variables. As we did in the last example, if k is an integer we regard $P(S_n \geq k)$ as $P(S_n \geq k - 0.5)$ and $P(S_n \leq k)$ as $P(S_n \leq k + 0.5)$. More generally, we replace each integer k in the set of interest by the interval $[k - 0.5, k + 0.5]$.

Example 6.22

Ipolito versus Power. In Example 2.40, we asked the question: what is the probability 101 disputed votes would change the outcome of an election in which the winner won by 17 votes. To do this they would have to split 59–42 or worse. In Chapter 2, we analyzed a similar situation by supposing that the 101 votes were sampled without replacement from the 2,827 cast. Here, we simplify by supposing that the votes were determined by flipping 101 coins.

Using notation and computations from the previous example, we are interested in

$$P(S_{101} \geq 58.5) = P\left(\frac{S_{101} - 50.5}{\sqrt{101}/2} \geq \frac{8}{\sqrt{101}/2}\right)$$
$$\approx P(\chi \geq 1.59) = 1 - 0.9441 = 0.0559$$

We leave it to the reader to decide if a 5.6% probability of changing the outcome justifies calling for a new election.

The next example shows that the histogram correction is a device not only to get more accurate estimates, but also allows us to get answers in cases where a naive application of the central limit theorem would give a senseless answer.

Example 6.23 Suppose we flip 16 coins. Use (6.17) to estimate the probability that we get exactly 8 heads.

The mean number of heads is $n/2 = 8$, while the standard deviation is $\sqrt{n}/2 = 2$. To use the normal approximation we write $\{S_{16} = 8\}$ as $\{7.5 \leq S_{16} \leq 8.5\}$. By (6.17),

$$P\left(\frac{7.5 - 8}{2} \leq \frac{S_{16} - 8}{2} \leq \frac{8.5 - 8}{2}\right) \approx P(-0.25 \leq \chi \leq 0.25)$$

Since $P(\chi = -0.25) = 0$, the probability of interest is $P(\chi \leq 0.25) - P(\chi \leq -0.25)$. The table tells us that $P(\chi \leq 0.25) = 0.5987$. There are no negative numbers in the table, but the normal distribution is symmetric, so

$$P(\chi \leq -0.25) = P(\chi \geq 0.25) = 1 - P(\chi \leq 0.25) = 1 - 0.5987 = 0.4013$$

and we have

$$P(S_{16} = 8) \approx 0.5987 - 0.4013 = 0.1974$$

The exact answer is

$$2^{-16} \frac{16!}{8!\,8!} = 0.1964$$

Similar reasoning shows that

$$P(S_{16} = 8 + k) \approx P\left(\chi \leq \frac{k + 0.5}{2}\right) - P\left(\chi \leq \frac{k - 0.5}{2}\right)$$

As the next table shows, these probabilities are fairly close to the exact answers obtained from the binomial distribution.

k	8	9	10	11	12	13
Normal approximations	0.1974	0.1747	0.1209	0.0656	0.0279	0.0092
Exact answers	0.1964	0.1746	0.1221	0.0666	0.0277	0.0085

Records of teams in the National Football League regular season could be modeled as the outcome of 16 coin tosses, one for each game. The next table compares the won–loss records of the 32 teams in the 2008 season with the numbers just computed:

k	8	9	10	11	12	13
$32 \times$ exact answers	6.28	5.59	3.91	2.13	0.88	0.27
Teams winning k games	5	6	1	4	4	1
Teams losing k games		2	1	2	3	0

Missing from the table are two teams who won only 2 games (an event with coin flip probability $120/2^{16} = 0.00183 \approx 1/546$) and the Detroit Lions who went $0 - 16$, an event with coin flip probability $1/65{,}536$.

Example 6.24 Suppose we roll a die 24 times. What is the probability that the sum of the numbers $S_{24} \geq 100$?

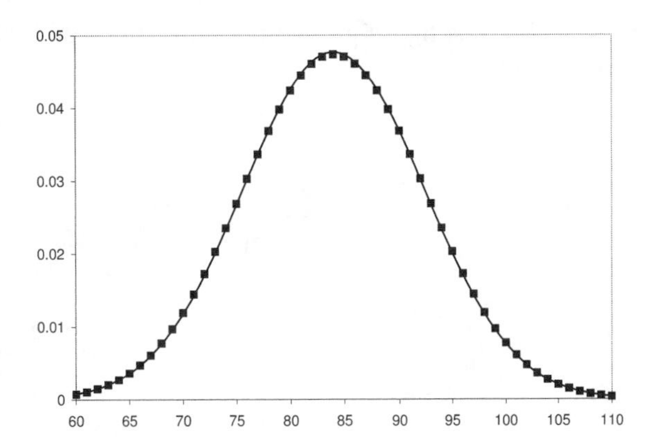

As the figure shows the distribution (squares) is approximately normal (smooth curve). To apply (6.17), the first step is to compute the mean and variance of S_{24}. $E\,S_{24} = 24 \cdot 7/2 = 84$. $\text{var}\,(S_{24}) = 24 \cdot 35/12 = 70$, so the standard deviation is $\sqrt{70} = 8.366$. Using (6.17) now, we have

$$P(S_{24} \geq 99.5) = P\left(\frac{S_{24} - 84}{\sqrt{70}} \geq \frac{15.5}{8.366}\right)$$
$$\approx P(\chi \geq 1.85) = 1 - P(\chi \leq 1.85) = 0.0322$$

compared with the exact answer 0.031760.

z-score. The key to finding the solution of this and the previous two problems is to compute the number of standard deviations separating the observed value from the expected value. The z-score is defined by

$$z = \frac{\text{observed value} - \text{expected value}}{\text{standard deviation}}$$

In the preceding example, this is

$$z = \frac{99.5 - 85}{8.366} = 1.85$$

so the normal approximation is $P(\chi \geq 1.85)$. Here a picture is worth a hundred words.

```
     Mean     1.85 s.d.
 -----+---------[///////---
      84        99.5
```

Example 6.25 **Roulette.** Consider a person playing roulette 1,000 times and betting \$1 on black each time. What is the probability that their net winnings are ≥ 0?

The figure gives three simulations of 1,000 plays of roulette, betting \$1 on black.

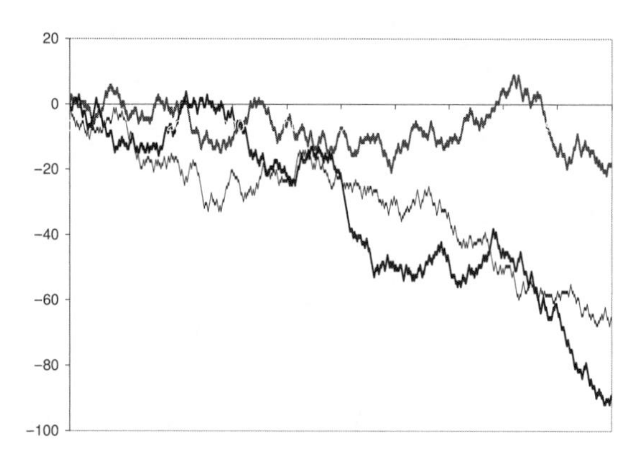

The outcome of the ith play $P(X_i = 1) = 18/38$, and $P(X_i = -1) = 20/38$, so $EX_i = -1/19 = -0.05263$, $EX_i^2 = 1$, and $\text{var}(X_i) = 1 - (1/19)^2 \approx 1$. The mean of 1,000 plays is -52.63, while the standard deviation is $\sqrt{1{,}000} = 31.62$. Writing ≥ 0 as ≥ -1 since only even numbers are possible values for $S_{1,000}$, the z-score is

$$z = \frac{-1 - (-52.63)}{31.62} = 1.63$$

so the normal approximation is $P(\chi \geq 1.63) = 1 - P(\chi \leq 1.63) = 0.0516$.

Mean 1.63 s.d.

−52.63 −1

Example 6.26

Normal approximation to the Poisson. Each year in Mythica, an average of 25 letter carriers are bitten by dogs. In the past year, 33 incidents were reported. Is this number exceptionally high?

Assuming that dog bites are a rare event, we use the Poisson distribution for the number of dog bites. As we observed in Example 6.3, a Poisson with mean 25 is the sum of 25 independent Poisson mean 1 random variables, so we can use the normal to approximate the Poisson. As the next figure shows the approximation is good but the Poisson still shows a slight skew.

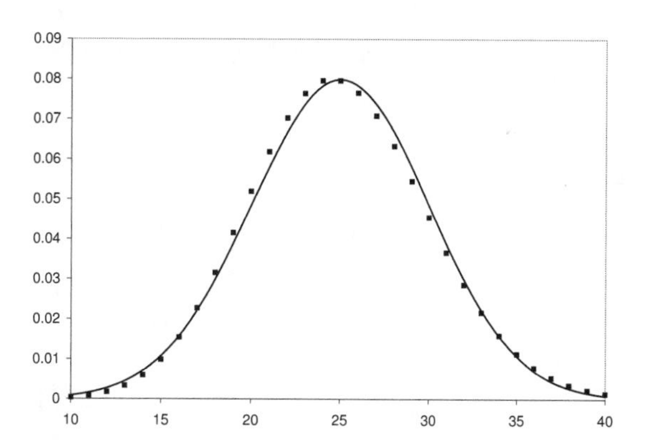

The mean is 25, while the standard deviation is $\sqrt{25} = 5$. Writing the observed event as ≥ 32.5 we see that this is $(32.5 - 25)/5 = 1.5$ standard deviations above the mean, so the normal approximation is $P(\chi \geq 1.25) = 1 - 0.8943 = 0.1057$, so this is not a very unusual event.

Example 6.27

A manufacturing plant produces boxes of biscuit mix that are 1 pound (454 grams). However, due to the poor flow properties of the powder the standard deviation of the weight of a box is 10 grams. A sample of 25 boxes had an average weight of 449.4 grams. Does this indicate a problem with the manufacturing process?

The average weight $\bar{X}_{25}$ has mean 454 grams and standard deviation $\sigma/\sqrt{n} = 10/\sqrt{25} = 2$. The observed weight is 4.6 grams below the mean, which is 2.3 standard deviations. From the Normal Table, $P(\chi \geq 2.3) = 1 - 0.9893 = 0.0107$, so a deviation this large by chance is rather unlikely.

Note. Continuous random variables can take on any value so there is no "histogram correction."

Example 6.28

Suppose that the average weight of a person is 182 pounds with a standard deviation of 40 pounds. A large plane can hold 400 people. What is the probability that the total weight of the people, S_{400}, will be more than 75,000 pounds?

The expected value of S_{400} is $400 \cdot 182 = 72{,}800$. The standard deviation is $\sigma\sqrt{n} = 40 \cdot 20 = 800$.

$$P(S_{400} \geq 75{,}000) = P\left(\frac{S_{420} - 72{,}800}{800} \geq 2.75\right)$$

$$\approx P(\chi \geq 2.75) = 1 - P(\chi \leq 2.75) = 0.003$$

In designing airplanes one cannot afford to make a mistake 3 times out of a 1,000 if the error will have disastrous consequences like a crash. Our table stops at 3.09. For larger values one can use the following approximation:

$$\int_x^\infty e^{-y^2/2}\,dy \leq \frac{1}{x}e^{-x^2/2} \tag{6.18}$$

Proof. Multiplying by y/x, which is ≥ 1 when $y \geq x$, we have

$$\int_x^\infty e^{-y^2/2}\,dy \leq \frac{1}{x}\int_x^\infty ye^{-y^2/2}\,dy$$

$$= \frac{1}{x}\left(-e^{-y^2/2}\right)\Big|_x^\infty = \frac{1}{x}e^{-x^2/2}$$

which proves the desired estimate. □

From this we see that the probability that the total weight is more than 77,600 pounds, which is six standard deviations above the mean, is at most

$$\frac{1}{\sqrt{2\pi}} \cdot \frac{1}{6}e^{-18} \approx 1 \times 10^{-9}$$

6.6 Applications to statistics

The central limit theorem is the key to several topics in statistics. In this section, we briefly discuss its use in hypothesis testing and confidence intervals, emphasizing the relevance of the central limit theorem and ignoring some of the fine points of the proper implementation of these methods (caveat emptor).

In the first situation we have a hypothesis about how the world works and ask if the data are consistent with that hypothesis.

Example 6.29 **Is there a difference between baseball and flipping coins?** In 2007, the Boston Red Sox won 96 games and lost 66. How likely is this result if the games were decided by flipping coins?

If they were flipping coins the mean number of wins would be $162/2 = 81$, the variance $162/4 = 40.5$, and standard deviation $\sqrt{40.5} = 6.36396$. The event $W \geq 96$ translates to $W \geq 95.5$ when we apply the histogram correction. This is 14.5 above the mean, or $14.5/6.36396 = 2.28$ standard deviations. The normal approximation for this probability is $P(\chi \geq 2.28) = 1 - 0.9887 = 0.0113$.

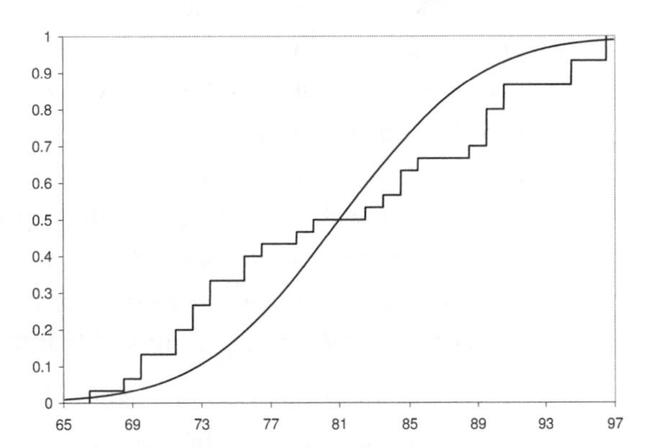

This probability is small but we must also remember that the Boston Red Sox are 1 of 30 teams in baseball ($1/30 = 0.033$) and we picked them because they had the best record. The figure compares the outcomes of the 2007 season with the normal distribution that we would have if games were decided by flipping coins. To be precise, it compares the number of teams with $\leq k$ wins with the distribution function of the approximating normal. Since the empirical distribution is less steep than the normal, we see that there is more variability in baseball win–loss records compared to flipping coins.

One final small point is that due to the histogram correction, we put the data point for k at $k + 0.5$. To see why we do this, note that the normal approximation for ≤ 91 is not 1/2, but is $P(\chi \leq 0.5/6.36396) = P(\chi \leq 0.08) = 0.5319$.

To delve further into the details of hypothesis testing, we will consider an artificial example.

Example 6.30

A factory owner is concerned that the fraction of red M&M's produced by his machines is not 1/4. In a sample of 1,200, 328 were red. Should he doubt the accuracy of his machines?

If the true fraction was $p = 1/4$, then in a sample of 1,200 candies the mean number of reds would be $1{,}200/4 = 300$ and the standard deviation would be $\sqrt{1{,}200(0.25)(0.75)} = 15$. ≥ 328 corresponds to $27.5/15 = 1.9$ standard deviations, so the observed event corresponds to $P(\chi \geq 1.83) = 0.0336$.

In this computation, we have ignored one of the subtleties referred to at the beginning of the section. If the factory owner had no suspicion of error before the test then he would have been equally alarmed by observing 272 reds, so we need to double the probability to 0.0672 to give a more accurate estimate of a deviation this large by chance alone.

Example 6.31

Weldon's dice data. An English biologist named Weldon was interested in the "pip effect" in dice – the idea that the spots, or "pips," which on some dice are produced by cutting small holes in the surface, make the sides with more spots lighter and more likely to turn up. Weldon threw 12 dice 26,306 times for a total of 315,672 throws and observed that a 5 or 6 came up on 106,602 throws or with probability 0.33770. Is this significantly different from 1/3?

If the true value was 1/3, then the expected number of successes would be 105,224, the variance 87,686.666, and the standard deviation 296.12. Ignoring the histogram correction the excess of 5's and 6's is 1,378 or 4.6535 standard deviations, so we are very confident that the true probability is not 1/3.

To make a statement about what we think the real value of p is, we note that if the true probability is p, then the average in the sample $\hat{p} = 106{,}602/315{,}672$, then $\hat{p}$ has mean p, and standard deviation $\sqrt{p(1-p)/n}$. Consulting the Normal Table we see that

$$P(-2 \leq \chi \leq 2) = 2(P(\chi \leq 2) - 1/2) = 0.9544$$

Replacing p by the estimate $\hat{p}$ in the formula for the standard deviation we see that 95% of the time the true value of p will lie in

$$\left[\hat{p} - \frac{2\sqrt{\hat{p}(1-\hat{p})}}{\sqrt{n}}, \quad \hat{p} + \frac{2\sqrt{\hat{p}(1-\hat{p})}}{\sqrt{n}}\right] \tag{6.19}$$

Plugging in our estimate $\hat{p}$, we have

$$0.33770 \pm 2\sqrt{\frac{0.3377 \cdot 0.6623}{315{,}672}} = 0.33770 \pm 0.00168$$

This is easier to understand if we subtract 1/3.

$$\hat{p} - 1/3 = 0.00427 \pm 0.00168 = [0.00259, 0.00595]$$

This difference, which is $\approx 1/250$, is not enough to be noticeable by people who play dice games for amusement, but is, perhaps, large enough to be of concern for a casino that entertains tens of thousands of gamblers a year and offers a wide variety of bets on dice games. For this reason most casinos use dice with no pips.

Example 6.32

Buffon's needle. As shown in Example 5.22 if the needle is 1/2 the distance between the cracks on the floor then the probability of hitting a crack is $1/\pi = 0.318310$. In three simulations of 10,000 trials, we observed 3,095, 3,186, and 3,237 hits. To see that these are "typical" outcomes, note that by (6.19) 95% of the observations will fall in

$$0.31831 \pm \frac{2\sqrt{(0.31831)(0.68169)}}{100} = 0.31831 \pm 0.00932 = [0.3090, 0.3276]$$

Example 6.33

Sample size selection. Suppose you want to forecast the outcome of an election and you are trying to figure out how many people to survey so that with probability 0.95 your guess does not differ from the true answer by more than 2%.

From (6.19) we see that if a fraction $\hat{p}$ of the people in a sample of size n are for candidate B then the **95% confidence interval** will be

$$\hat{p} \pm \frac{2\sqrt{\hat{p}(1-\hat{p})}}{\sqrt{n}}$$

To get rid of the $\hat{p}$'s in the width of the confidence interval we note that the function $g(x) = x(1-x) = x - x^2$ has derivative $g'(x) = 1 - 2x$, which is positive for $x < 1/2$ and negative for $x > 1/2$. So g is increasing for $x < 1/2$ and decreasing for $x > 1/2$, and hence the maximum value occurs at $x = 1/2$. Noticing that $2\sqrt{x(1-x)} = 1$ when $x = 1/2$, we have

$$P\left(p \in \left[\hat{p} - \frac{1}{\sqrt{n}}, \hat{p} + \frac{1}{\sqrt{n}}\right]\right) \geq 0.95 \qquad (6.20)$$

To see that this approximation is reasonable for elections, notice that if $0.4 \leq p \leq 0.6$, then $\sqrt{p(1-p)} \geq \sqrt{0.24} = 0.4899$, compared with our upper bound of 1/2. Even when p is as small as 0.2, $\sqrt{p(1-p)} = \sqrt{0.16} = 0.4$.

To answer our original question now, we set $1/\sqrt{n} = 0.02$ and solve to get

$$n = (1/0.02)^2 = 50^2 = 2{,}500$$

To get the error down to 1%, we would need $n = 1/(0.01)^2 = 10{,}000$. Comparing the last two results and noticing that the radius of the confidence interval in (6.20) is $1/\sqrt{n}$, we see that to reduce the error by a factor of 2 requires a sample that is $2^2 = 4$ times as large.

Example 6.34

The *Literary Digest* poll. In order to forecast the outcome of the 1936 election, *Literary Digest* polled 2.4 million people and found that 57% of them were going to vote for Alf Landon and 43% were going to vote for Franklin Roosevelt. A 95% confidence interval for the true fraction of people voting for Landon based on this sample would be 0.57 ± 0.00064, but Roosevelt won, getting 62% of the vote to Landon's 38%.

To explain how this happened we have to look at the methods *Literary Digest* used. They sent 10 million questionnaires to people whose names came from telephone books and club membership lists. Since many of the 9 million unemployed did not belong to clubs or have telephones the sample was not representative of the population as a whole. A second bias came from the fact that only 24% of the people filled out the form. This problem was mentioned in our discussion of exit polls in Example 3.19. If, for example, 36% of Landon voters and 16.6% of Roosevelt voters responded then the fraction of people who responded would be $0.62(0.166) + 0.38(0.36) = 0.24$ and the fraction in the sample for Landon would be

$$\frac{0.38(0.36)}{0.62(0.166) + 0.38(0.36)} = \frac{0.1368}{0.24} = 0.57$$

in agreement with the data.

Finally, we would like to observe that *Literary Digest*, which soon after went bankrupt, could have saved a lot of money by taking a smaller sample. George Gallup, who was just then getting started in the polling business, predicted based on a sample of size 50,000 that Roosevelt would get 56% of the vote. His 95% confidence interval for the election result would be 0.56 ± 0.0045, compared with the election result of 62%. Again there could be some bias in his sample, or perhaps Landon voters, discouraged by the predicted outcome, were less likely to vote. The moral of our story is: It is much better to take a good sample than a large one.

6.7 Exercises

Expected value of sums

1. A man plays roulette and bets \$1 on black 19 times. He wins \$1 with probability 18/38 and loses \$1 with probability 20/38. What are his expected winnings?

2. Suppose we draw 13 cards out of a deck of 52. What is the expected value of the number of aces we get?

3. Suppose we pick 3 students at random from a class with 10 boys and 15 girls. Let X be the number of boys selected and Y be the number of girls selected. Find $E(X - Y)$.

4. 12 ducks fly overhead. Each of 6 hunters picks one duck at random to aim at and kills it with probability 0.6. (a) What is the mean number of ducks that are killed? (b) What is the expected number of hunters who hit the duck they aim at?

5. 10 people get on an elevator on the first floor of a seven-story building. Each gets off at one of the six higher floors chosen at random. What is the expected number of stops the elevator makes?

6. Suppose Noah started with n pairs of animals on the ark and m of them died. If we suppose that fate chose the m animals at random, what is the expected number of complete pairs that are left?

7. Suppose we draw 5 cards out of a deck of 52. What is the expected number of different suits in our hand? For example, if we draw K♠ 3♠ 10♡ 8♡ 6♣, there are three different suits in our hand.

8. Suppose we draw cards out of a deck without replacement. How many cards do we expect to draw out before we get an ace? Hint: The locations of the four aces in the deck divide it into five groups $X_1, \ldots, X_5$.

9. A calculus class has 150 students. Assume there are 365 days in a year and note that $150/365 = 0.410959$. (a) What is the probability that at least one student is born on April 1? (b) Let N be the number of days on which at least one student has a birthday. Find the expected value EN.

Variance and covariance

10. Roll two dice and let $Z = XY$ be the product of the two numbers obtained. What is the mean and variance of Z?

11. Suppose X and Y are independent with $EX = 1$, $EY = 2$, $\text{var}(X) = 3$, and $\text{var}(Y) = 1$. Find the mean and variance of $3X + 4Y - 5$.

12. In a class with 18 boys and 12 girls, boys have probability 1/3 of knowing the answer and girls have probability 1/2 of knowing the answer to a typical question the teacher asks. Assuming that whether or not the students know the answer are independent events, find the mean and variance of the number of students who know the answer.

13. At a local high school, 12 boys and 4 girls are applying to MIT. Suppose that the boys have a 10% chance of getting in and the girls have a 20% chance. (a) Find the mean and variance of the number of students accepted. (b) What is more likely: 2 boys and no girls accepted or 1 boy and 1 girl?

14. Let N_k be the number of independent trials we need to get k successes when success has probability p. Find the mean and variance of N_k.

15. Suppose we roll a die repeatedly until we see each number at least once and let R be the number of rolls required. Find the mean and variance of R.

16. Suppose X takes on the values $-2, -1, 0, 1, 2$ with probability 1/5 each, and let $Y = X^2$. (a) Find $\text{cov}(X, Y)$. (b) Are X and Y independent?

Chebyshev's inequality

17. Suppose that it is known that the number of items produced at a factory per week is a random variable X with mean 50. (a) What can we say about the probability $X \geq 75$? (b) Suppose that the variance of X is 25. What can we say about $P(40 < X < 60)$?

18. Let $X = \text{binomial}(4, 1/2)$. Use Chebyshev's inequality to estimate $P(|X - 2| \geq 2)$ and compare with the exact probability.

19. Let $\bar{X}_{10,000}$ be the fraction of heads in 10,000 tosses. Use Chebyshev's inequality to bound $P(|\bar{X}_n - 1/2| \geq 0.01)$ and the normal approximation to estimate this probability.

20. Let X have a Poisson distribution with mean 16. Estimate $P(X \geq 28)$ using (a) Chebyshev's inequality and (b) the normal approximation.

Central limit theorem, I. Coin flips

21. A person bets you that in 100 tosses of a fair coin the number of heads will differ from 50 by 4 or more. What is the probability that you will win this bet?

22. Suppose we toss a coin 100 times. Which is bigger, the probability of exactly 50 heads or at least 60 heads?

23. Bill is a student at Cornell. In any given course he gets an A with probability 1/2 and a B with probability 1/2. Suppose the outcomes of his courses are independent. In his 4 years at Cornell he will take 33 courses. If he can get 22 A's and only 11 B's he can graduate with a 3.666 average. What is the probability that he will do this?

24. In a 162-game season find the approximate probability that a team with a 0.5 chance of winning will win at least 87 games.

25. British Airways and United offer identical service on two flights from New York to London that leave at the same time. Suppose that they are competing for the same pool of 400 customers who choose an airline at random. What is the probability that United will have more customers than its 230 seats?

26. A probability class has 30 students. As part of an assignment, each student tosses a coin 200 times and records the number of heads. What is the probability that no student gets exactly 100 heads?

27. A fair coin is tossed 2,500 times. Find a number m so that the chance that the number of heads is between $1{,}250 - m$ and $1{,}250 + m$ is approximately 2/3.

CLT, II. Biased coins

28. Suppose we roll a die 600 times. What is the approximate probability that the number of 1's obtained lies between 90 and 110?

29. Suppose that each of 300 patients has a probability of 1/3 of being helped by a treatment. Find approximately the probability that more than 120 patients are helped by the treatment.

30. Suppose that 10% of a certain brand of jelly beans are red. Use the normal approximation to estimate the probability that in a bag of 400 jelly beans there are at least 45 red ones.

31. A basketball player makes 80% of his free throws on the average. Use the normal approximation to compute the probability that in 25 attempts he will make at least 23.

32. Suppose that we roll two dice 180 times and we are interested in the probability that we get exactly 5 double sixes. Find (a) the normal approximation, (b) the exact answer, and (c) the Poisson approximation.

33. A gymnast has a difficult trick with a 10% chance of success. She tries the trick 25 times and wants to know the probability that she will get exactly two successes. Compute the (a) exact answer, (b) Poisson approximation, and (c) normal approximation.

34. A student is taking a true/false test with 48 questions. (a) Suppose she has a probability $p = 3/4$ of getting each question right. What is the probability that she will get at least 38 right? (b) Answer the last question if she knows the answers to half the questions and flips a coin to answer the other half. Notice that in each case the expected number of questions she gets right is 36.

35. To estimate the percent of voters who oppose a certain ballot measure, a survey organization takes a random sample of 200 voters. If 45% of the voters oppose the measure, estimate the chance that (a) exactly 90 voters in the sample oppose the measure and (b) more than half the voters in the sample oppose the measure.

36. An airline knows that in the long run only 90% of passengers who book a seat show up for their flight. On a particular flight with 300 seats there are 324 reservations. Assuming passengers make independent decisions what is the chance that the flight will be overbooked?

37. Suppose that 15% of people don't show up for a flight, and suppose that their decisions are independent. How many tickets can you sell for a plane with 144 seats and be 99% sure that not too many people will show up?

38. A seed manufacturer sells seeds in packets of 50. Assume that each seed germinates with probability 0.99 independently of all the others. The manufacturer promises to replace, at no cost to the buyer, any packet with 3 or more seeds that do not germinate. (a) Use the Poisson to estimate the probability a packet must be replaced. (b) Use the normal to estimate the probability that the manufacturer has to replace more than 70 of the last 4,000 packets sold.

39. An electronics company produces devices that work properly 95% of the time. The new devices are shipped in boxes of 400. The company wants to guarantee that k or more devices per box work. What is the largest k so that at least 95% of the boxes meet the warranty?

CLT, III. General distributions

40. The number of students who enroll in a psychology class is Poisson with mean 100. If the enrollment is >120, then the class will be split into two sections. Estimate the probability that this will occur.

41. On each bet a gambler loses \$1 with probability 0.7, loses \$2 with probability 0.2, and wins \$10 with probability 0.1. Estimate the probability that the gambler will be losing after 100 bets.

42. Suppose we roll a die 10 times. What is the approximate probability that the sum of the numbers obtained lies between 30 and 40?

43. Members of the Beta Upsilon Zeta fraternity each drink a random number of beers with mean 6 and standard deviation 3. If there are 81 fraternity members, how much should they buy so that using the normal approximation they are 93.32% sure they will not run out?

44. An insurance company has 10,000 automobile policy holders. The expected yearly claim per policy holder is \$240 with a standard deviation of \$800. Approximate the probability that the yearly claim exceeds \$2.7 million.

45. A die is rolled repeatedly until the sum of the numbers obtained is larger than 200. What is the probability that you need more than 66 rolls to do this?

46. Suppose that the checkout time at a grocery store has a mean of 5 minutes and a standard deviation of 2 minutes. Estimate the probability that a checker will serve at least 49 customers during her 4-hour shift.

Confidence intervals

47. Of the first 10,000 votes cast in an election, 5,180 were for candidate A. Find a 95% confidence interval for the fraction of votes that candidate A will receive.

48. A bank examines the records of 150 patrons and finds that 63 have savings accounts. Find a 95% confidence interval for the fraction of people with savings accounts.

49. Among 625 randomly chosen Swedish citizens, it was found that 25 had previously been citizens of another country. Find a 95% confidence interval for the true proportion.

50. A sample of 2,809 handheld video games revealed that 212 broke within the first 3 months of operation. Find a 95% confidence interval for the true proportion that break in the first 3 months.

51. Suppose we take a poll of 2,500 people. What percentage should the leader have for us to be 99% confident that the leader will be the winner?

52. For a class project, you are supposed to take a poll to forecast the outcome of an election. How many people do you have to ask so that with probability 0.95 your estimate will not differ from the true outcome by more than 5%?

Hypothesis testing

We will use casual language to state these problems. The precise formulation of the first one is if $p = 1/6$, then what is the probability that we will observe 3,123 or more sixes in 18,000 rolls. Here, and in what we follows, we will ignore the fact that we should multiply by 2 to get a more accurate idea of how odd the observation is.

53. A casino owner is concerned based on past experience that his dice show 6 too often. He makes his employees roll a die 18,000 times and they observe 3,123 sixes. Is the die biased?

54. We suspect that a bridge player is cheating by putting an honor card (ace, king, queen, or jack) at the bottom of the deck when he shuffles so that this card will end up in his hand. In 16 times when he dealt, the last card dealt to him was an honor on 9 occasions. Are we confident that he is cheating?

55. If both parents carry one dominant (A) and one recessive gene (a) for a trait then Mendelian inheritance predicts that 1/4 of the offspring will have both recessive genes (aa) and show the recessive trait. If among 96 offspring of Aa parents we find 30 are aa, is this consistent with Mendelian inheritance?

56. A softball player brags that he is a 0.300 hitter, yet at the end of the season he has gotten 21 hits in 84 at bats. Is this just bad luck?

57. In a 60-day period in Ithaca 12 days were rainy. Is this observation consistent with the belief that the true proportion of rainy days is 1/3?

58. In a poll of 900 Americans in 1978, 65% said that extramarital sex was wrong, whereas a similar poll in 1985 found that 72% had the same opinion. Are we confident that opinions have changed?

59. A psychic claims to be able to guess the suit of a card without seeing it. In 52 attempts, someone who is just guessing will get 13 right on the average. How many would he have to get right so that we are about 99.88% sure he is not guessing.

7

Option Pricing

7.1 Discrete time

In the next section we discuss the Black–Scholes formula. To prepare for that, we consider the much simpler problem of pricing options when there are a finite number of time periods and two possible outcomes at each stage. The restriction to two outcomes is not as bad as one might think. One justification for this is that we are looking at the process on a very slow timescale, so at most one interesting event happens (or not) per time period. We begin by considering a very simple special case.

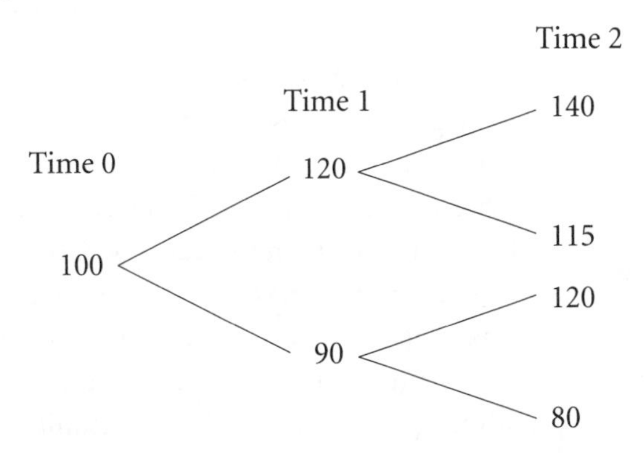

Example 7.1 **Two-period binary tree.** Suppose that a stock price starts at 100 at time 0. At time 1 (one day or one month or 1 year later) it is either worth 120 or 90. If the stock is worth 120 at time 1, then it might be worth 140 or 115 at time 2. If the price is 90 at time 1, then the possibilities at time 2 are 120 and 80. Suppose now that you are offered a **European call option** with **strike price** 100 and **expiry** 2. This means you have an option to buy the stock (but not an obligation to do so) for 100 at time 2, that is, after seeing the outcome of the first and second stages. If the stock price is 80, you do not exercise the option to purchase the stock and your profit is 0. In the other cases you choose to buy the stock at 100 and then immediately sell it at X_2 to get a payoff of $X_2 - 100$, where X_2 is the stock price at time 2. Combining the two cases we can write the payoff in

general as $(X_2 - 100)^+$, where $z^+ = \max\{z, 0\}$ denotes the positive part of z. Our problem is to figure out what is the right price for this option.

At first glance this may seem impossible since we have not assigned probabilities to the various events. However, it is a miracle of "pricing by the absence of arbitrage" that in this case we do not have to assign probabilities to the events to compute the price. To explain this we start by considering a small piece of the tree. When $X_1 = 90$, X_2 will be 120 ("up") or 80 ("down") for a profit of 30 or a loss of 10, respectively. If we pay c for the option then when X_2 is up we make a profit of $20 - c$, but when it is down we make $-c$. The last two sentences are summarized in the following table:

	Stock	Option
Up	30	$20 - c$
Down	-10	$-c$

Suppose we buy x units of the stock and y units of the option, where negative numbers indicate that we sold instead of bought. One possible strategy is to choose x and y so that the outcome is the same if the stock goes up or down:

$$30x + (20 - c)y = -10x + (-c)y$$

Solving, we have $40x + 20y = 0$ or $y = -2x$. Plugging this choice of y into the last equation shows that our profit will be $(-10 + 2c)x$. If $c > 5$, then we can make a large profit with no risk by buying large amounts of the stock and selling twice as many options. Of course, if $c < 5$, we can make a large profit by doing the reverse. Thus, in this case the only sensible price for the option is 5.

A scheme that makes money without any possibility of a loss is called an **arbitrage opportunity**. It is reasonable to think that these will not exist in financial markets (or at least be short-lived), since if and when they exist people take advantage of them and the opportunity goes away. Using our new terminology we can say that the only price for the option that is consistent with absence of arbitrage is $c = 5$, so that must be the price of the option (at time 1 when $X_1 = 90$).

Before we try to tackle the whole tree to figure out the price of the option at time 0, it is useful to look at things in a different way. Generalizing our example, let $a_{i,j}$ be the profit for the ith security when the jth outcome occurs.

Theorem 7.1. *Exactly one of the following holds:*

(i) There is a betting scheme $x = (x_1, x_2, \ldots, x_n)$ so that $\sum_{i=1}^m x_i a_{i,j} \geq 0$ for each j and $\sum_{i=1}^m x_i a_{i,k} > 0$ for some k.

(ii) There is a probability vector $p = (p_1, p_2, \ldots, p_n)$ with $p_j > 0$ so that $\sum_{j=1}^n a_{i,j} p_j = 0$ for all i.

Here a vector x satisfying (i) is an arbitrage opportunity. We never lose any money but for at least one outcome we gain a positive amount. Turning to (ii), the vector p is called a martingale measure since if the probability of the jth outcome is p_j, then the expected change in the price of the ith stock is equal to 0. Combining the two interpretations we can restate Theorem 7.1 as

Theorem 7.2. *There is no arbitrage if and only if there is a strictly positive probability vector so that all the stock prices are martingale.*

Why is this true? One direction is easy. If (i) is true, then for any strictly positive probability vector $\sum_{i=1}^m \sum_{j=1}^n x_i a_{i,j} p_j > 0$, so (ii) is false.

Suppose now that (i) is false. The linear combinations $\sum_{i=1}^m x_i a_{i,j}$ when viewed as vectors indexed by j form a linear subspace of n-dimensional Euclidean space. Call it $\mathcal{L}$. If (i) is false, this subspace intersects the positive orthant $\mathcal{O} = \{y: y_j \geq 0 \text{ for all } j\}$ only at the origin. By linear algebra we know that $\mathcal{L}$ can be extended to an $(n-1)$-dimensional subspace $\mathcal{H}$ that only intersects $\mathcal{O}$ at the origin.

Since $\mathcal{H}$ has dimension $n-1$, it can be written as $\mathcal{H} = \{y : \sum_{j=1}^n y_j p_j = 0\}$. Since for each fixed i the vector $a_{i,j}$ is in $\mathcal{L} \subset \mathcal{H}$, (ii) holds. To see that all the $p_j > 0$, we leave it to the reader to check that if not, there would be a nonzero vector in $\mathcal{O}$ that would be in $\mathcal{H}$. □

To apply Theorem 7.1 to our simplified example we begin by noting that in this case $a_{i,j}$ is given by

		$j=1$	$j=2$
Stock	$i=1$	30	-10
Option	$i=2$	$20-c$	$-c$

By Theorem 7.2 if there is no arbitrage, then there must be an assignment of probabilities p_j so that

$$30p_1 - 10p_2 = 0 \qquad (20-c)p_1 + (-c)p_2 = 0$$

From the first equation we conclude that $p_1 = 1/4$ and $p_2 = 3/4$. Rewriting the second we have

$$c = 20p_1 = 20 \cdot (1/4) = 5$$

To generalize from the last calculation to finish our example we note that the equation $30p_1 - 10p_2 = 0$ says that under p_j the stock price is a martingale (that is, the average value of the change in price is 0), while $c = 20p_1 + 0p_2$ says that the price of the option is then the expected value under the martingale probabilities. Using these ideas we can quickly complete the computations in our example. When $X_1 = 120$, the two possible scenarios lead to a change of +20 or −5, so the relative probabilities of these two events should be 1/5 and 4/5. When $X_0 = 100$, the possible price changes on the first step are +20 and −10, so their relative probabilities are 1/3 and 2/3. Drawing a picture of the possibilities, we have

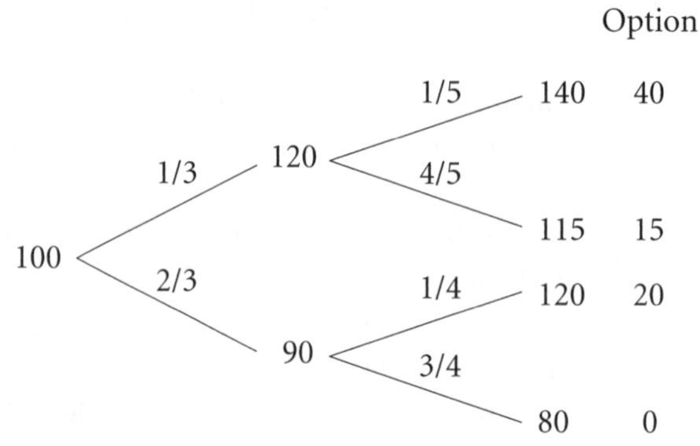

so the value of the option is

$$\frac{1}{15} \cdot 40 + \frac{4}{15} \cdot 15 + \frac{1}{6} \cdot 20 = \frac{80 + 120 + 100}{30} = 10 \qquad \square$$

The last derivation may seem a little devious, so we now give a second derivation of the price of the option. In the scenario described above, our investor has four possible actions:

A_0. Put \$1 in the bank and end up with \$1 in all possible scenarios.
A_1. Buy one share of stock at time 0 and sell it at time 1.
A_2. Buy one share at time 1 if the stock is at 120 and sell it at time 2.
A_3. Buy one share at time 1 if the stock is at 90 and sell it at time 2.

These actions produce the following payoffs in the indicated outcomes:

Time 1	Time 2	A_0	A_1	A_2	A_3	Option
120	140	1	20	20	0	40
120	115	1	20	−5	0	15
90	120	1	−10	0	30	20
90	80	1	−10	0	−10	0

Noting that the payoffs from the four actions are themselves vectors in four-dimensional space, it is natural to think that by using a linear combination of these actions we can reproduce the option exactly. To find the coefficients we write four equations in four unknowns:

$$\begin{aligned} z_0 + 20z_1 + 20z_2 &= 40 \\ z_0 + 20z_1 - 5z_2 &= 15 \\ z_0 - 10z_1 + 30z_3 &= 20 \\ z_0 - 10z_1 - 10z_3 &= 0 \end{aligned} \tag{7.1}$$

Subtracting the second equation from the first and the fourth from the third gives $25z_2 = 25$ and $40z_3 = 20$, so $z_2 = 1$ and $z_3 = 1/2$. Plugging in these values, we have two equations in two unknowns:

$$z_0 + 20z_1 = 20 \qquad z_0 - 10z_1 = 5$$

Taking differences, we conclude that $30z_1 = 15$, so $z_1 = 1/2$ and $z_0 = 10$.

The reader may have already noticed that $z_0 = 10$ is the option price. This is no accident. What we have shown is that with \$10 cash we can buy and sell shares of stock to produce the outcome of the option in all cases. In the terminology of Wall Street, $z_1 = 1/2$, $z_2 = 1$, $z_3 = 1/2$ is a **hedging strategy** that allows us to **replicate the option.** Once we can do this it follows that the fair price must be \$10. To do this note that if we could sell it for \$12, then we can take \$10 of the cash to replicate the option and have a sure profit of \$2.

7.2 Continuous time

To do option pricing in continuous time we need a model of the stock price, and for this we have to first explain **Brownian motion.** Let $X_1, X_2, \ldots,$ be independent and take the values 1 and -1 with probability 1/2 each. $EX = 0$ and $EX^2 = 1$, so if we let $S_n = X_1 + \cdots + X_n$, then $S_n/\sqrt{n}$ converges to χ a standard normal distribution. Intuitively, Brownian motion is what results when we look not only at time n but also at how the process got there. To be precise, we let $t \geq 0$ and consider $S_{[nt]}/\sqrt{n}$, where $[nt]$ is the largest integer $\leq nt$. In words, we multiply n by t and then round down to the nearest whole number. When $n = 1{,}000$ the picture looks like

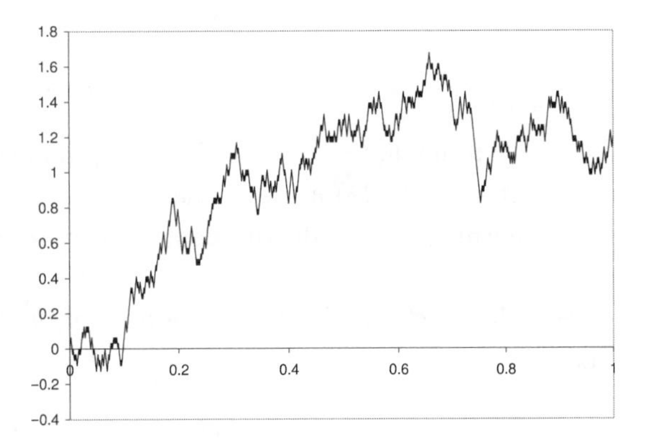

To understand the nature of the limit process we note that

$$\frac{S_{[nt]}}{\sqrt{n}} = \frac{S_{[nt]}}{\sqrt{[nt]}} \cdot \frac{\sqrt{[nt]}}{\sqrt{n}}$$

The first term approaches a standard normal distribution and the second $\sqrt{t}$, so $S_n/\sqrt{n}$ converges to $\sqrt{t}\chi$, a normal with mean 0 and variance t. Repeating the reasoning in the last paragraph we can see that if $s < t$, then $(S_{[nt]} - S_{[ns]})/\sqrt{n}$ converges to a normal with mean zero and variance $t - s$. Noting that $(S_{[nt]} - S_{[ns]})$ is independent of $S_{[ns]}$ suggests the following definition of the limiting process that we call **Brownian motion**.

- B_t has a normal distribution with mean 0 and variance t.
- If $0 < t_1 < \cdots < t_n$ then $B_{t_1}, B_{t_2} - B_{t_1}, \ldots, B_{t_n} - B_{t_{n-1}}$ are independent

In modeling stock prices it is natural to assume that the daily percentage changes in the price are independent. For this reason and the mundane fact that stock prices must be >0, we model the stock as what is called **geometric Brownian motion.**

$$X_t = X_0 \cdot \exp(\mu t + \sigma B_t) \tag{7.2}$$

μ is the exponential growth rate of the stock and σ its volatility. In writing the model we have assumed that the growth rate and volatility of the stock are constant. If we also assume that the interest rate r is constant, then the discounted stock price is

$$e^{-rt} X_t = X_0 \cdot \exp((\mu - r)t + \sigma B_t)$$

Here, we have to multiply by e^{-rt}, since \$1 at time t has the same value as e^{-rt} dollars today.

Our problem is to determine the fair price of a European call option $(X_t - K)^+$ with strike price K and expiry t. Extrapolating wildly from Theorem 7.2, we can say that any consistent set of prices must come from a martingale measure. This implies

$$\mu = r - \sigma^2/2 \tag{7.3}$$

To compute the value of the call option, we need to compute its value in the model in (7.2) for this special value of μ. Using the fact that $\log(X_t/X_0)$ has a normal$(\mu t, \sigma^2 t)$ distribution, one can show

Black–Scholes formula. *The price of the European call option $(X_T - K)^+$ is given by*

$$X_0 \Phi(\sigma\sqrt{t} - \alpha) - e^{-rt} K \Phi(-\alpha)$$

where Φ is the distribution function of a standard normal and

$$\alpha = \{\log(K/X_0 e^{\mu t})\}/\sigma\sqrt{t}$$

To try to come to grips with this ugly formula note that $K/X_0e^{\mu t}$ is the ratio of the strike price to the expected value of the stock at time t under the martingale probabilities, while $\sigma\sqrt{t}$ is the standard deviation of $\log(X_t/X_0)$.

Example 7.2

Microsoft call options. The February 23, 1998, *Wall Street Journal* listed the following prices for July call options on Microsoft stock.

Strike	75	80	85
Price	11	$8\frac{1}{8}$	$5\frac{1}{2}$

On this date Microsoft stock was trading at $81\frac{5}{8}$, while the annual interest rate was about 4% per year. Should you buy the call option with strike 80?

Solution. The answer to this question depends on your opinion of the volatility of the market over the period. Suppose that we follow a traditional rule of thumb and decide that $\sigma = 0.3$; that is, over a 1-year period a stock's price might change by about 30% of its current value. In this case the drift rate for the martingale measure is

$$\mu = r - \sigma^2/2 = 0.04 - (0.09)/2 = 0.04 - 0.045 = -0.005$$

and so the log ratio is

$$\log(K/X_0e^{\mu t}) = \log(80/(81.625e^{-0.005(5/12)}) = \log(80/81.455) = -0.018026$$

Five months corresponds to $t = 5/12$, so the standard deviation

$$\sigma\sqrt{t} = 0.3\sqrt{5/12} = 0.19364$$

and $\alpha = -0.018026/0.19364 = -0.09309$. Plugging in now, we have a price of

$$\begin{aligned}
&81.625\Phi(0.19365 + 0.09309) - e^{-0.04(5/12)}80\Phi(0.09309)\\
&= 81.625\Phi(0.28674) - 78.678\Phi(0.09309)\\
&= 81.625(0.6128) - 78.678(0.5371) = 50.02 - 42.25 = 7.76
\end{aligned}$$

This is somewhat lower than the price quoted in the paper. There are two reasons for this. First, the options listed in the *Wall Street Journal* are

American call options. The holder has the right to exercise at any time during the life of the option. Since one can ignore the additional freedom to exercise early, American options are at least as valuable as their European counterparts. Second, and perhaps more importantly, we have not spent much effort on our estimate of r and σ. Nonetheless, as the next example shows the predictions of the formula are in rough agreement with the observed process.

Example 7.3

Intel call options. Again consulting the *Wall Street Journal* for February 23, 1998, we find the following prices listed for July call options on Intel stock, which was trading at $94\frac{3}{16}$.

Strike	70	75	80	85	90	95	100	105
Price	26	22	18	$14\frac{1}{2}$	$11\frac{3}{8}$	$8\frac{3}{4}$	$6\frac{1}{2}$	$4\frac{3}{8}$
Formula	25.65	21.16	17.01	13.59	10.11	7.11	5.39	4.13

Normal Table

$$\Phi(x) = \int_{-\infty}^{x} \frac{1}{\sqrt{2\pi}} e^{-y^2/2} \, dy$$

To illustrate the use of the table: $\Phi(0.36) = 0.6406$, $\Phi(1.34) = 0.9099$

	0	1	2	3	4	5	6	7	8	9
0.0	0.5000	0.5040	0.5080	0.5120	0.5160	0.5199	0.5239	0.5279	0.5319	0.5359
0.1	0.5398	0.5438	0.5478	0.5517	0.5557	0.5596	0.5636	0.5675	0.5714	0.5753
0.2	0.5793	0.5832	0.5871	0.5910	0.5948	0.5987	0.6026	0.6064	0.6103	0.6141
0.3	0.6179	0.6217	0.6255	0.6293	0.6331	0.6368	0.6406	0.6443	0.6480	0.6517
0.4	0.6554	0.6591	0.6628	0.6664	0.6700	0.6736	0.6772	0.6808	0.6844	0.6879
0.5	0.6915	0.6950	0.6985	0.7019	0.7054	0.7088	0.7123	0.7157	0.7190	0.7224
0.6	0.7257	0.7291	0.7324	0.7357	0.7389	0.7422	0.7454	0.7486	0.7517	0.7549
0.7	0.7580	0.7611	0.7642	0.7673	0.7703	0.7734	0.7764	0.7793	0.7823	0.7852
0.8	0.7881	0.7910	0.7939	0.7967	0.7995	0.8023	0.8051	0.8078	0.8106	0.8133
0.9	0.8159	0.8186	0.8212	0.8238	0.8264	0.8289	0.8315	0.8340	0.8365	0.8389
1.0	0.8413	0.8438	0.8461	0.8485	0.8508	0.8531	0.8554	0.8577	0.8599	0.8621
1.1	0.8643	0.8665	0.8686	0.8708	0.8729	0.8749	0.8770	0.8790	0.8810	0.8830
1.2	0.8849	0.8869	0.8888	0.8907	0.8925	0.8943	0.8962	0.8980	0.8997	0.9015
1.3	0.9032	0.9049	0.9066	0.9082	0.9099	0.9115	0.9131	0.9147	0.9162	0.9177
1.4	0.9192	0.9207	0.9222	0.9236	0.9251	0.9265	0.9279	0.9292	0.9306	0.9319
1.5	0.9332	0.9345	0.9357	0.9370	0.9382	0.9394	0.9406	0.9418	0.9429	0.9441
1.6	0.9452	0.9463	0.9474	0.9484	0.9495	0.9505	0.9515	0.9525	0.9535	0.9545
1.7	0.9554	0.9564	0.9573	0.9582	0.9591	0.9599	0.9608	0.9616	0.9625	0.9633
1.8	0.9641	0.9649	0.9656	0.9664	0.9671	0.9678	0.9686	0.9693	0.9699	0.9706
1.9	0.9713	0.9719	0.9726	0.9732	0.9738	0.9744	0.9750	0.9756	0.9761	0.9767
2.0	0.9772	0.9778	0.9783	0.9788	0.9793	0.9798	0.9803	0.9808	0.9812	0.9817
2.1	0.9821	0.9826	0.9830	0.9834	0.9838	0.9842	0.9846	0.9850	0.9854	0.9857
2.2	0.9861	0.9864	0.9868	0.9871	0.9875	0.9878	0.9881	0.9884	0.9887	0.9890
2.3	0.9893	0.9896	0.9898	0.9901	0.9904	0.9906	0.9909	0.9911	0.9913	0.9916
2.4	0.9918	0.9920	0.9922	0.9924	0.9927	0.9929	0.9931	0.9932	0.9934	0.9936
2.5	0.9938	0.9940	0.9941	0.9943	0.9945	0.9946	0.9948	0.9949	0.9951	0.9952
2.6	0.9953	0.9955	0.9956	0.9957	0.9959	0.9960	0.9961	0.9962	0.9963	0.9964
2.7	0.9965	0.9966	0.9967	0.9968	0.9969	0.9970	0.9971	0.9972	0.9973	0.9974
2.8	0.9974	0.9975	0.9976	0.9977	0.9977	0.9978	0.9979	0.9979	0.9980	0.9981
2.9	0.9981	0.9982	0.9982	0.9983	0.9984	0.9984	0.9985	0.9985	0.9986	0.9986
3.0	0.9986	0.9987	0.9987	0.9988	0.9988	0.9989	0.9989	0.9989	0.9990	0.9990

Answers to Odd Problems

1. Basic Concepts

1. $\Omega = \{ABC, ACB, BAC, BCA, CAB, CBA\}$, 3. 1/4, 5. (a) 16/36, (b) 20/36, 7. 3/4, 9. 1:4/57, 2:16/57, 3:15/57, 4:12/57, 11. Number of outcomes out of 216: (a) 1, (b) 3, (c) 6, (d) 10, (e) 15, (f) 21, (g) 25, (h) 27, 13. 90, 15. 77, 17. 0.2, 19. Yes, 21. Yes, 25. 0.44, 27. 19/27, 29. (a) 0.048, (b) 0.296, (c) 0.464, (d) 0.192, 31. 0.4032, 33. 0.4914, 35. 0:6/36, 1:10/36, 2:8/36, 3:6/36, 4:4/36, 5:2/36, 37. The same as for the sum of two ordinary dice, 39. 7, 41. 1/4, 43. $-17/216$, 45. Flag or Joker: $-13/54$, 20: $-12/54$, 10: $-10/54$, 5: $-12/54$, 2: $-9/54$, 1: $-6/54$, 47. 32/10, 49. 26, 51. mean 5.8125, variance 1.03, 53. Mean 3.888, variance 0.432. 55. No.

2. Combinatorial Probability

1. 9!, 3. 3,360, 5. (a) 151,200, (b) 210, 7. 55, 9. (a) $26^3 10^3$ (b) 0.6391, 11. 0.0605, 13. (a) 540, (b) 372, 15. 118,800, 17. (a) 240, (b) 480, 19. $(P_{15,5})^4 P_{15,4}$, 21. (a) 120, (b) 60, (c) 504,000, (d) 34,650, 23. (a) 50!/(8!7!), (b) $C_{50,8} C_{50,7}$, 25. 103,680, 27. (a) 4!=24, (b) 576, (c) 105, (d) 2,520, (e) 40,320, (f) 384, 29. 60/1024, 31. (a) 0.152, (b) 0.618, 33. (a) 0.05469, (b) 0.92981, 35. 21/32, 37. (a) 0.3874, (b) 0.3773, (c) 0.3725, 39. 0.3297, 41. Poisson: 0.778801, Exact: 0.778703, 43. 0.1889, 45. 0.9735, 47. 0.4232, 49. 0.11192, 51. 0.6664, 53. 10/21. 55. 0.3343, 57. $C_{26,5} C_{4,1} / C_{30,6}$, 59. 0.3684, 61. (a) 0.4085, (b) 0.5107, (c) 0.0766, 63. $C_{12,5} \cdot C_{10,5} \cdot 5!$, 65. 5.47×10^{-4}, 67. 1/2, 69. 0.5814, 71. 0.2285, 73. (a) 0.7969, (b) 0.1992, (c) 0.0038, 75. (a) Five of a kind: 0.000771, (b) four of a kind: 0.019290, (c) full house: 0.038580, (d) three of a kind: 0.154320, (e) two pair: 0.231481, (f) one pair: 0.462962, (g) no pair: 0.092592, 77. 5/32, 79. (a) 0.9, (b) 0.3, 81. 0.3439, 83. 0.028, 85. $P(A) = 0.1655$, $P(A)/P(B) = 3.9971$, 87. First bound: 0.1667, second: 0.1551, third: 0.1555213, exact: 0.1555124

3. Conditional Probability

1. 2/3, 3. 4/25, 5. (a) 0.4385, (b) 0.4561, (c) 0.1052, 7. 6/7, 9. 2/3, 11. 1/3, 13. $P(A \cap B) = 1/8$, $P(B) = 3/8$, $P(A \cup B) = 1/2$, 15. 0.6, 17. Put 1 white ball in the first urn and all the other balls in the second. 19. 11/216, 21. 14: 146/1296,

23. 0.9568, 25. 0.17, 27. 2/5, 29. 0.7, 31. 0.4918, 33. 1/4, 1/3, 35. 1/51, 37. (a) 0.9, (b) 1/730, 39. 12/29, 41. 8/9, 43. 56/65, 45. 4/5, 47. 1/7, 49. (a) 1/2, (b) 5/9, 51. 8/9, 53. 5/6, 55. 4/7, 57. 4/7, 2/7, 1/7, 59. 0.062, 0.3387, 61. $P(N_1 = k, N_6 = j) = P(N_1 = j, N_6 = k) = (4/6)^{j-1}(5/6)^{k-j-1}(1/6)^2$, 63. (a) 1/6 (b) 1/2,

65.

Y	X=1	2	3
1	1/9	2/9	1/3
2	1/18	0	1/6
3	0	1/9	0

4. Markov Chains

1. (a) $x = 0.4$, $y = 0.4$, $z = 0.6$, (b) $x = 0.2$, $y = 0.5$, $z = 0.2$, 3. 0.294, 5. 13/32, 7. (b) 0.22, 0.166, 9. 51.2% in 1980, 56.8% in 1990, and 60.9% in 2000, 11. (a) 0.55, (b) 0.575, (c) 0.6, 13. 0.7825, 0.727291, $8/11 = 0.0727272$, 15. 38%, 25%, 8/33 = 24%, 17. 0.211, 0.286, 0.502, 19. 4/19, 21. (b) 7/13, 23. 4/5, 25. 0.25, 0.5, 0.25, 27. (a) 0.35, 0.34, 0.31, (b) 0.5294, 0.3235, 0.1470, 29. Long run frequencies are A:0.2, C:0.55, T:0.25, 31. (b) 0.2321, 0.1964, 0.5714, 33. 0.2817, 35. $\pi(0) = 100/122$, $\pi(1) = 10/122$, $\pi(2) = 2/122$, $\pi(3) = 10/122$, 37. $\pi(0) = 0.1, \pi(1) = 0.4, \pi(2) = 0.3, \pi(3) = 0.2$, 39. 1/3, 41. 16.666, 43. (a) 1/4, (b) 3.916, 45. 0.125/0.685

5. Continuous Distributions

1. $c = 1/3$, 3. (a) 1/2, (b) 0.3, (c) 0.05, 5. (a) 15/28, (b) 127/49, (c) 2.304, 7. Yes. $x^{-2}e^{-1/x}$, 9. (a) $x^2/4$ for $0 \le x \le 2$, (b) 1/4, (c) 7/16, (d) $\sqrt{2}$, 11. (a) $x^{1/2}$ for $0 \le x \le 1$, (b) $1 - 1/\sqrt{3}$, (c) 1/6, (d) 1/4, 13. any number in [0, 1], 15. (a) $1 - x^{-3}$ (b) $(1-u)^{-1/3}$, 21. $y^{-1/2}/2$, 23. $1 - \exp(-e^{-x})$, 25. (a) $f(y) + f(-y)$ for $0 \le y \le 1$, (b) $\{f(\sqrt{z}) + f(-\sqrt{z})\}/2\sqrt{z}$ for $0 \le z \le 1$, 27. (a) $c = 1$, (b) 3/8, 29. $(1-z)^2/2$ for $0 \le z \le 1$, 31. 7/16, 33. $F(x, y) = \min\{x, y\}$ if $x, y > 0$ and $\min\{x, y\} \le 1$, $F(x, y) = 1$ if $\min\{x, y\} > 1$, 0 otherwise, 35. $F_X(x) = F_{X,Y}(x, \infty) = \lim_{y\to\infty} F_{X,Y}(x, y)$, 37. (a) $f_X(x) = (2/\pi)\sqrt{1-x^2}$, (b) $f_Y(y|X = x) = 1/2\sqrt{1-x^2}$ for $-\sqrt{1-x^2} < y < \sqrt{1-x^2}$, 39. $f_X(x) = 3(1-x)^2$ for $0 < x < 1$, $f_Y(y) = 6y(1-y)$ for $0 < y < 1$, (b) $f_X(x|Y = y) = 1/(1-y), 0 < x < 1-y$

6. Limit Theorems

1. −1, 3. −3/5, 5. 5.03, 7. 3.114, 9. 123.136, 11. Mean 6, variance 43, 13. (a) mean 2, variance 1.82, (b) 1 boy and 1 girl, 15. Mean 14.7, variance 38.99, 17. (a) $P(X \ge 75) \le 2/3$, (b) $P(40 < X < 60) \ge 0.75$, 19. Chebyshev: $\le 1/4$, normal approx: 0.0455, 21. 0.484, 23. 0.041, 25. 0.0016, 27. 24, 29. 0.0085, 31. 0.1056, 33. (a) 0.2658, (b) 0.2565, (c) 0.2475, 35. (a) 0.0566, (b) 0.0884, 37. 157, 39. 373, 41. 0.3897, 43. 527, 45. 0.0116, 47. [0.508, 0.528], 49. [0.0044, 0.0356], 51. 52.33, 53. 0.0081, 55. Yes, 57. No, 59. 23

Index of Terms

Index of Examples